CRITICAL CULTURAL COMMUNICATION

General Editors: Sarah Banet-Weiser and Kent A. Ono

Dangerous Curves: Latina Bodies in the Media
Isabel Molina-Guzmán

The Net Effect: Romanticism, Capitalism, and the Internet
Thomas Streeter

T0336186

William Blake's 1795/circa 1805 print "Newton," represents the scientist Isaac Newton in a way that expresses Blake's view of the limits of the calculated scientific reasoning for which Newton was famous. The colors and texture of the rock and the body loom over the bright but small page of measurement in the lower right hand corner, expressing Blake's belief in the primacy of the creative imagination. Or as Blake wrote himself in *The Marriage of Heaven and Hell*, "What is now proved was once, only imagin'd."

The Net Effect

Romanticism, Capitalism, and the Internet

Thomas Streeter

NEW YORK UNIVERSITY PRESS
New York and London

NEW YORK UNIVERSITY PRESS
New York and London
www.nyupress.org

References to Internet websites (URLs) were accurate at the time of writing.
Neither the author nor New York University Press is responsible for URLs
that may have expired or changed since the manuscript was prepared.

Library of Congress Cataloging-in-Publication Data
Streeter, Thomas.
The net effect : romanticism, capitalism, and the internet / Thomas Streeter.
p. cm. — (Critical cultural communication)
Includes bibliographical references and index.
ISBN 978–0–8147–4115–3 (cl : alk. paper) — ISBN 978–0–8147–4116–0
(pb : alk. paper) — ISBN 978–0–8147–4117–7 (ebook)
1. Computers and civilization. 2. Computers—Social aspects.
3. Information technology—Social aspects. 4. Internet—Social aspects.
I. Title.
QA76.9.C66S884   2010
303.48'33—dc22   2010024294

New York University Press books are printed on acid-free paper,
and their binding materials are chosen for strength and durability.
We strive to use environmentally responsible suppliers and materials
to the greatest extent possible in publishing our books.

Manufactured in the United States of America
c   10 9 8 7 6 5 4 3 2 1
p   10 9 8 7 6 5 4 3 2 1

To my childhood friends The Hacks, with whom I discovered the boyish pleasures of technological fiddling long ago and who understood hacking before the term had been layered with political connotations. I thank them for their dedicated friendship over the decades.

Contents

Acknowledgments

THIS WORK WAS long in coming, as I grappled with a moving target while adjusting to life's ordinary surprises. Here I can only mention some of the individuals who helped me along the way. Lisa Henderson gave me excellent critical readings, discussions, patience, and most of all wonderful affirmation while this book finally came together. Thanks to Sylvia Schafer for suggesting the title and providing encouragement at a time when I badly needed it. Excellent advice, criticism, and intelligent discussion came from many, including Michael Curtin, Christina Dunbar-Hester, Kathy Fox, Tarleton Gillespie, Mary Lou Kete, Beth Mintz, John Durham Peters, Christian Sandvig, Ross Thomson, and Fred Turner. Thanks to Ben Peters and Rasmus Kleis Nielsen for their enthusiasm and for inviting me to some stimulating seminars. Thanks to the principals of the Key Centre for Cultural Policy Studies in Brisbane, Australia, for a stimulating fellowship in the summer of 1999. I am grateful to the Institute for Advanced Study for support during 2000–2001 and to the faculty, especially Clifford Geertz and Joan Scott, and all the members of the School of Social Science that year for smart and helpful comments, criticism, conversation, and encouragement. Thanks to series editors Sarah Banet-Weiser and Kent A. Ono, to NYU Press editors, reviewers, and staff, and to copyeditor Jay Williams for their excellent help, suggestions, and tolerance of my foibles. I am grateful to the faculty union of the University of Vermont, United Academics, which soaked up my time but also provided me a window onto what a mature approach to democratic decision making might actually look like. And last but by no means least, thanks to my son Seth, who makes life a constant surprise and a joy, no matter how dark his music.

Introduction

"Communication" is a registry of modern longings.

— John Durham Peters[1]

IT IS STILL common in some circles to assume that rationality, technology, and the modern are somehow opposed to or fundamentally different from culture, the imagination, nature, and expression. This book starts from the premise that this is not so and that the internet is prima facie evidence of that. The internet has been tangled up with all manner of human longings, in both obvious ways—for example, the internet stock bubble—and more subtle ways, such as certain aspects of its technical design and trends in its regulation. In hopes of better understanding both technology and longings, this book gives that entanglement a close look.

Part of what emerges in looking at the internet this way is our networked desktop computers are not so much direct descendants of the giant computers of the 1960s as they are reactions against those computers and what they represented, a reaction that was to some degree cultural. Beginning in the 1960s, engineers who had different impulses for how to build and use computers began to draw on what is properly called romanticism to construct justifications for their alternative designs. By the 1970s and 1980s, skilled popular writers like Stewart Brand, Ted Nelson, and Steven Levy joined them to elaborate these gestures into a more fully articulate vision.

The original giant computers were often associated with misguided efforts to somehow calculate our way out of human dilemmas: to control the horror of nuclear warfare, for example, or to win the Vietnam war, or to industrialize secretarial work, or to turn school children into studious and obedient users of electronic encyclopedias. Sensing the folly of these plans to use computers to control human complexity and to frame it in a predictable grid, increasing numbers of individuals began to reinterpret the act of computing as a form of expression, exploration, or art, to see themselves as artist, rebel, or both, and to find communities with similar experiences that would reinforce that interpretation. People need to express themselves, it was said, people want and need spontaneity, creativity, or dragon-slaying heroism, and direct, unplanned interaction with computers offered a kind of enticing, safely limited unpredictability that would fulfill those goals. That is why we need small computers instead of mainframes, the argument went, why we need personal computers instead of dedicated word processors, why we need the open, end-to-end distributed networking of the internet instead of proprietary corporate systems, why we should invest in 1990s dotcoms, why we need open source software. These discursive habits, I have found, had consequences. For example, neoliberalism's quarter century reign as a hegemonic political economic ideology owes much to the linkage of romantic tropes to networked computing. At the same time, the internet has become an important collective thought object for considering new ways of thinking about democracy.

None of this is caused by romanticism alone. Causes are complicated, and in any case romanticism, as I understand it, is always a reaction *to* something; it is in the specific dynamics of its interaction with other trends, we will see, that romanticism can have consequences. But what this book suggests is that the specific forms of the life-shaping digital machinery we have surrounded ourselves with are not the product of some kind of technological necessity; it is not that we once had a mistaken idea about what computers were for and now have discovered their "true" uses. Nor is this "the market" at work alone; most of what is described in this book takes place in situations where buying and selling are not the operative forces. The point is that, while economic and technological forces of course have played a role, the internet's construction is peppered with profoundly cultural forces: the deep weight of the remembered past and the related, collectively organized pressures of human passions made articulate.

This is a book, then, about America's romance with computer communication, a history of the dense interaction of the American social and political imagination with the development of internet technology. It is a look at how culture has influenced the construction of the internet and how the structure of the internet has played a role in cultures of social and political thought. In that sense, it is a case study in "how institutions think."[2] *The Net Effect* explores various ways computer communication has been conceived over the years of its development, with a focus on conceptions that have influenced policy. Beginning with the 1950s, when computers were primarily imagined as machines for rapidly solving complex mathematical problems, the book traces the appearance and character of other notions of what connected computers might be for: as means for fighting nuclear wars in the 1950s, for example, as systems for bringing mathematical certainty to the messy complexity of social life in the early 1960s, as automated writing and reading machines for enlightening individuals in the late 1960s, as countercultural playgrounds in the 1970s, as an icon for what's good about free markets in the 1980s, as a new frontier to be conquered in the early 1990s, and, by the late 1990s, as the transcendence of markets in an anarchist open source utopia. The book is not just about the truthfulness of these various conceptions—inaccuracies are often revealing—but about their *effects* accurate or not, their impact both on the construction of the internet and on its reception in other parts of life.

Approach:
How the Feel of Modern Life Shapes Modern Living

Instead of looking at the internet as a harbinger of the future, *The Net Effect* looks at it more as an expression of the times. This is not a book about the road ahead, inventing the future, the next big thing, or the future of ideas, creativity, or the economy. Nor is it a warning about what might happen or what might be lost if we do not act. Sometimes exploring the complexity of what has actually happened offers more useful insight than the urgent gropings of prognostication. So, rather like Walter Benjamin's angel of history, *The Net Effect* looks backwards more than forwards.[3] It focuses as much on ways that social and cultural trends have shaped the internet as on how the internet has shaped trends, and it finds the imprint of themes from the preinternet past in places where others have seen sharp historical breaks. It does all this by mixing historical storytelling with discussions of philosophical and theoretical issues.[4] And it is written with a sense of inquiry, with more of an eye towards answering questions than winning arguments.

This book began, then, with several sets of questions. One set came out of my earlier work. In *Selling the Air* I found that the development of broadcast technology—easily as mind-blowing in 1920 as the internet was in 1994—can be seen as a kind of social philosophy in practice, as something that was as much a product of social visions as of technical or economic necessities. Over the long term, I found that broadcast policy was neither a blueprint for reality nor just an ideology that legitimates or enables decisions made elsewhere. Rather, policy's contradictions and misrecognitions were themselves a key part of the social construction of the institutions and technologies of broadcasting; the focus was on the productivity of policy discourses, even when they were contradictory.[5] As the internet grew in shape and force in the 1990s, I was struck by the parallels between the 1920s and the 1990s and wondered how the visions associated with the internet might similarly be shaping policymaking.

As I watched developments with these parallels in mind, however, I was struck by two more things: first, the remarkable revival of the market-enamored political economic practices of neoliberalism in the mid-1990s and, second, the often noted but not fully explained extent to which something as dry and seemingly technocratic as computer network policymaking was riddled with odd moments of passion, often in ways that seemed to confound the received ideas about the nature of corporate capitalism. Beginning with an essay first published in 1999,[6] I sought to develop an explanation of how rebellion, self-expression, and technology and market policies seemed to be harnessed together in a historically unique way, including in places where one would least expect it, such as computing systems funded by the military.

And, the more I thought about all of these concerns, the more they seemed intertwined. Understanding one of them depended on understanding the others. So, finally, the book expanded into an exploration of how the feel of modern life shapes modern living, an inquiry into the interactions of subjectivities or personal experiences with technological, political, and economic relations. My initial observations about the internet became the basis for a case study that helped understand larger questions about culture, society, and modern life.[7]

How do broadly shared habits of thought change over time? Some writers work through the history of ideas, as read through the lives and writings of famous authors. We inherit our ideas about rights, liberty, and markets from John Locke and Adam Smith, it is said, or the role of the 'sixties counterculture in computing in the nineties can be understood by a close look at the life and work of Stewart Brand, whose influential career spanned both periods. Others look more to culture and find zeitgeists or worldviews in cultural forms. Jacob Burckhardt saw a Renaissance spirit in the art and architecture of sixteenth-century Italy, for example, and more recently scholars have seen postmodern celebrations of the malleable self in the cyberpunk-influenced advertisements, novels, and films of the 1980s.[8] While I have borrowed from work in both these traditions, my own approach tackles issues on a more sociological level.

Traditional intellectual history tends to carefully trace ideas over time through the biographies of individuals who take up those ideas and assumes that the ideas have meaning and coherence through those biographies. This has the advantage of linking the development of ideas to real individuals and their direct contacts with others; it is an approach that eschews overgeneralization or a hand-waving approach to ideologies. Yet locating the coherence of a system of thought in the biographies of individuals also risks a false clarity. John Locke articulated an individualist theory of property rights, but the analogies between what he wrote and the intellectual habits of "possessive individualism" central to Western capitalism do not explain the popularity of the idea or why his theories of rights and property are referenced but his views of religion are as often as not ignored.[9] Stewart Brand's ideas from the 1960s were indeed carried into the cyberculture in the 1980s and 1990s, but that does not explain why that importation was successful or why some aspects of his work got attention in the 1960s (for example, environmentalism, a distaste for the singular pursuit of wealth) and others in the 1990s (for example, computer technologies and a libertarian inclination towards markets). There are cases where famous authors in the field of computing sometimes changed their minds or said things that in retrospect seem incoherent or irrelevant. Similarly, drawing broad conclusions about society at large from films, novels, and advertisements risks assuming too much. Does Apple's *1984* TV ad for the Macintosh computer, broadcast nationally only once, tell us about the culture at large in the 1980s, or just about a small subset of that culture?

A century's worth of scholarship in the sociology of knowledge suggests a few principles for understanding the place of ideas in social life. First, ideas do not exist as isolated bits that can be picked up and discarded separately. Rather, they live and die insofar as they are sustained by their place in broad patterns of thought, in paradigms, in systems of value and belief that provide general visions of the world. (The main limitation of Richard Dawkin's popular notion of "memes" is precisely that it treats ideas as singular bits, as if they existed autonomously from larger systems of thought.)[10] When digital pioneer Douglas Engelbart first proposed in the 1960s that computers might be controlled interactively by a keyboard and a mouse through a windowing interface, this was not just the invention of a few devices. Engelbart was a key figure in a movement that was considering a completely different picture of what computers were about. It was a vision of how computers could be distributed communication tools—that is, something much like we understand them today—instead of the 1960s notion of computers as centralized and centralizing calculation and management devices. Engelbart presented an alternative worldview of computing, a different system of thought, of which the mouse and overlapping windows were simply expressions. If you only look at the mouse or the interface in isolation, you miss the underlying vision that made them possible.

Second, ideas emerge within communities. There are unique individuals who make important contributions, but those contributions generally grow out of, and are nurtured within, communities that share a system of thought or inquiry. Isaac Newton discovered calculus, but it is hardly a coincidence that Leibniz came up with the same ideas at roughly the same time.[11] Engelbart's ideas would have gone nowhere without a community of the like-minded, or at least the receptive, around him. Hence, the first thing to look for is shifts in the shared broad patterns, in what is in the air at a given time. And the principal objects of analysis are communities who share ideas, knowledge, methods, and habits of thought and talk. Individuals' actions are most important when they express the character of and changes within broader systems of thought.

Third, ideas inevitably exist in relations to social structures—complex relations, to be sure, but never completely autonomously. Ideas need living sustenance, that is, communities of people with resources and institutional relations that enable them to actively propagate and maintain themselves. A theology needs a church and a community of believers; a new approach to computer use requires a source of funding and an institutional home. But these connections are rarely formulaic. In the history of the internet, for every visionary like Engelbart there are other important figures with no explicit grand vision, political or otherwise. Andries Van Dam, for example, sometimes credited as one of the three pioneers of hypertext, is a modest college professor and researcher who approached computer programming with a spirit of cheerful professional craftsmanship rather than visionary ardor. While others prognosticated, he built working programs and, most importantly, taught generations of students about new possibilities for using computers, many of whom went on to key places in the industry. Beyond an enthusiasm for promoting computers as communication devices, his efforts show few overt signs of influence from cultural trends or from politics.

So the fact that some computer scientists sported long hair or wore antiwar pins as they built the internet does not make it inherently countercultural, just as the fact that their funding was largely from the military does not by itself make it a war machine. It is rare that systems of thought can be simply linked to broad social structures in mechanical, one-to-one fashion.[12] What priests tell their parishioners about birth control or divorce may be one thing, but what the community actually does may be another. The same may be said about an engineer's grant proposal that talks about using computers for military research, while his graduate students write protocols for email distribution lists that get used to discuss politics and science fiction novels. To say that institutions support ideas, therefore, is not to assume the existence of a clean, unbreakable link between official beliefs and the political valence of the machines that get built.

This messy area of connections between systems of thought and institutions remains a challenge. As we will see, new ideas often gain traction, not just because of an encounter with a big theory, but with small, everyday experiences: the compulsive draw that often comes with computer use, for example, or the repeated wonder of plugging in a new gizmo that a short time ago would have been impossibly expensive or just impossible, or the cubicle dweller's secret pleasure of discovering, on a slow day at work, something striking on computer networks that is unknown to the powers that be. Systems of thought often work at the level of tacit habits of talk and action rather than explicit belief systems and become visible through the accumulation of decisions over time. Outside the graduate seminar or the hard sciences, at least, changes in systems of thought seem to be as much about habits of the heart as habits of the mind. (The notion of "memes" may remain compelling in popular usage because, with its emphasis on things like buzzwords and slogans, it loosely captures the informal dynamic by which new ideas catch on: by slogans, passions, and implicit, culturally specific forms of "common sense," as much as by rational axioms, evidence, principles, and doctrines.) Sometimes it is more important that an idea be thrilling than it be logically compelling.

To the extent that this book offers a generalizable method, it is to tackle this microstructural problem of the interplay of ideas and institutions by looking at connections among three levels: (1) shared felt experiences associated with technologies; (2) cultural traditions that people draw on to make sense of those experiences; and (3) articulations between those linked traditions and experiences with political ideas, particularly political ideas that shape policymaking around internet structure.

People think with texts and theories, but in working on this book I found that they also think with objects and institutions. Whether or not computers themselves think, they are things that people think with, things that inspire us to think about our selves and our relations to others. Big ideas, like a revived belief in the justice of markets or an enthusiasm for digital democracy, are sometimes brought in to help individuals account for and connect their everyday experiences with machines to life as a whole rather than the other way around. Intellectual trends thus can gain traction starting at the level of everyday experience and only then drawing from more formally structured statements of principle. This book looks for the philosophy, then, not just in fully articulated theories, but at how ideas and everyday experiences of life in general and computers interact. Keeping an eye on how things might have felt from the bottom up, I found, sometimes better explained the success of writers' ideas than did the lives or works of the writers themselves.

Effects: The Net Effect Is in the Making of It

But how, when working from the bottom up, can one sort out the significant from the trivial? Is there any connection at all, for example, between the internet and the cold war military visions that underwrote the internet's early development and the technology? Or between the internet and the utopian democratic claims made by many internet pioneers?

Instead of working from texts to zeitgeists, I looked for occasions where cultural trends made a material difference. The book looks at instances where people draw on various cultural systems to make sense of common feelings associated with computers and then how those acts of making sense play roles in formal and informal policy making. I looked for cases, in other words, where the intersections of intellectual frameworks and feelings can be seen to actually shape policy decisions influencing the construction of the internet and its social instantiation, cases where changes in policymaking occurred that cannot be otherwise entirely accounted for.[13]

As I worked on this book, people would often assume a book called *The Net Effect* was studying the effect of the internet on people—on children, perhaps, or education, families, or nations. That is not exactly the question here. On the one hand, it is simply too soon to tell what the internet's effects are. True social change is long and deep, taking place over decades or centuries. Scholars are still debating, quite thoughtfully, the effect of the printed book on human civilization several centuries after its widespread adoption. The full effect of the century-old telephone remains something of a sociological mystery. Gauging the social effects of a brand new technology like the internet—as of this writing, barely more than a decade old as a consumer item, still changing almost monthly in its character and reach—is bound to be largely an exercise in guesswork and sloganeering.

On the other hand, there's a question of what one means by effects. Sociologists and historians of technology are quick to tell us to be wary of overly simple forms of technological determinism, in which a technology like television or the internet is imagined as if it were exterior to society, as if it dropped from the sky fully formed and then exerted effects on that society from the outside. Technologies, it is said, are socially constructed.[14] They are deeply embedded in and shaped by social processes and choices and so should not be thought of as something outside of or autonomous from society. This is particularly true of the internet. The choices that go into computer design are not purely technological; the same microprocessor, for example, can guide a missile, run a word processor, or power a home game console, and which of these gets implemented is at least to a large degree a social choice.[15] (And even if computers sometimes have unintended consequences, even if they surprise us, that surprise may be more about us than it is about anything inherent to the machines; consider the unexpected popularity of email in the early days of computer networking.) Contemporary computing, therefore, is in an important way the product of a gradual accumulation of social and cultural choices, choices among competing visions of computers' purposes and social capacities. These choices, in turn, typically rest on those collections of tacit assumptions that power social relations—assumptions about social hierarchy, for example, or constructions of self. As Donna Haraway once put it, technologies are "frozen moments of the fluid social interactions constituting them."[16] To the extent that this is true, then the interesting question is not, what is the effect of the internet on society? but, how has the internet been socially constructed and what role has that process of construction played in society?[17] What did we learn from the way the internet was built, from the unique way that it appeared and came into broad public consciousness?

Social constructionism, however, is more a way of framing the problem than a solution to it. To the extent that computers are simply, as Sherry Turkle's early work suggests, a Rorschach blot onto which we project our dreams and understandings, one can safely discount the specifics of the technology and just focus on how people imagine it.[18] The internet, certainly, has been frequently looked at through the lens of various utopias and described alternately as, say, the embodiment of the competitive free market or the embodiment of communitarian cooperation. Such claims are interesting, but in the first instance they generally tell us more about the political orientation of the claimants than they do about the internet.

But, on another level, the internet, more obviously than many other technologies, has been and continues to be a gradual, collective work in progress. The way it is built and organized is inseparable from the way its builders imagine it, even if they do imagine it partially or inaccurately. So the more important (and, analytically, more difficult) question is, how have various shared visions, even the inaccurate ones, shaped policymaking around the internet? How have they shaped its construction and, therefore, its character, its role in social life? How have culture and policy interacted to make the internet what it is?

This is the principal methodological question that drives this book. As I approached the massive, sprawling tangle of technical information, personal narratives, and political events that make up the history of the internet, I have sought out instances in which culture played a key role in broad policy and design choices. The internet has figured in many ways in culture—in movies, for example, or novels, or dating habits, even in religion—but I have pursued those instances where culture has demonstrably made a difference in the construction of the internet itself. This book's approach to the question of causality, then, is to understand the internet not as a thing that has an effect but as itself a process of social construction. The net effect is in the making of it.[19]

Culture, Selves, Power

Who are you when, on an ordinary day, you sit down to use a computer? Are you a citizen? A consumer? A manager? A technician? An artist? Are you looking for the familiar, or are you hoping to be surprised? Are you trying to reaffirm who you are, your sense of self? Or are you perhaps hoping to break out of your routine, to experience something different, a better self?

This book suggests that the different answers to these questions offered by culture, that is, shifting varieties of learned self-understanding or selfhood, have made a difference in the development of the internet and that the ways this has happened tells us something about the character of modern life. Multiple forms of self-understanding are at play at any one time; in the last half-century in the United States; for example, utilitarian and managerial constructs of the self have played a key role. But I also look at the role of the romantic self, where the self is understood as the source of a dynamic, inner experience that calls on us to live creatively beyond the bounds of predictable rationality. We are romantics even, and perhaps especially, in the face of high technologies.

From Locke through Burckhardt to Tocqueville to postmodernism, the question of how societies imagine the self is a recurring theme. In particular, the traditional history of ideas teaches the importance and deep complexities of the historical evolution of what Ian Watt called "that vast complex of interdependent factors denoted by the term 'individualism'"[20] and what poststructuralists suggested was the study of the process of the I in history. The idea here is not that the self is an illusion, nor that "society" mechanically determines our identities, nor that the self has suddenly become, in the postmodern era, infinitely malleable. Rather, as Christina Dunbar-Hester has argued, "the benefit of using [the category of identity] is to get at parts of human experience that are moving targets, slippery, constructed, and yet 'real.'"[21]

To get at the "slippery, constructed, yet real" character of subjectivity, I find it useful to follow John Frow, who has written of "the imaginary forms of selfhood through which we experience the world and our relation to it."[22] Forms of selfhood, in this sense, are forms, not types of individuals. They are discursive patterns embedded in institutions and historical processes that become available to individuals as ways of making sense of who they are in given contexts. One never simply *is* a utilitarian or romantic or gendered self. Rather, most of us find it necessary or useful to adopt roles, to think and speak of ourselves in various established ways, at various moments in our lives. We often have to think of ourselves, for example, as alternately passionate and as administrators, one moment as caring parents or partners and the next as self-interested rational actors in a marketplace and after that as competent professionals with resumes. "Imaginary forms of selfhood," then, are neither fixed identities nor complete or determinate in some kind of mechanical way. They are plural and fluid, but not infinitely so; there are typically several forms available to any given individual in any given context, and it is possible, and probably sometimes necessary, to move among them.[23] We all regularly negotiate the tensions inherent in this situation in our own ways, of course, but the contingencies of social process and history provide us a shifting set of available strategies for accomplishing that negotiation.[24]

Are there particular forms of selfhood associated with computing? There certainly has been speculation along those lines. Software engineer and *Wired* contributor Ellen Ullman, for example, has written evocatively about what she calls "a male sort of loneliness that adheres in programming." Yet she hints at the layers of complexity in the phenomenon when she quips, "Fifteen years of programming, and I've finally learned to take my loneliness like a man."[25]

One of the problems with some of the original work on the history of individualism was a tendency to imagine a singular, European or Western self, as if everyone in a given time and place experiences the world in the same way. Rhetoric about the spread of the internet and computing frequently echoes this conceit when it speaks in universal terms—"everyone" is on the internet, using email, using Facebook, and so on—in a way that systematically ignores cultural and economic barriers to access and differences in use.[26] For example, the percentage of women entering the fields of computer science has always been small and, according to some reports, has actually declined in the last decade. Because this has occurred at a time when women's participation in many other professions has been going up, most attribute this pattern to a mix of cultural, institutional, and economic barriers.[27]

In response to the blindness inherent in the tradition of a singular we, in the assertion of a unified and universal sense of self, there is now an established set of critiques. From W. E. B. Du Bois's *The Souls of Black Folk* through the feminist writings of the 1970s through the literatures of cultural studies today, the emphasis has been on the *different* forms of selfhood experienced by different groups, and the problems and pain caused by dominant groups' tendencies to imagine their own experiences of selfhood as the only experiences. (It is not always wrong to speak of a we in general; and when I do so, sometimes it signifies myself and the reader and sometimes it speaks of the shared future of humanity.)[28] But, in the end, there is no everyone on the internet, just as there is no single type of Western individual, and the tendency to speak as if there is, the tendency to speak of a we that encompasses everyone, is both inaccurate and potentially manipulative.

Yet the response cannot be only to say that everyone's experience is different or to assert other identities against, say, a white male identity. When Ullman describes "the male sort of loneliness that inheres in programming," she is both pointing to a generalizable pattern of experience our culture associates with a type of masculinity and allowing how, as a woman, she can share in that experience; after fifteen years of programming, she can take her loneliness like a man. The experience is associated with, but not necessarily bound to, masculinity, and the question is how that association has been historically constructed.

Cultural studies' project was never simply the liberal one of giving voice to the voiceless or of asserting one kind of experience mechanically tied to a social grouping against another; it was a rethinking of how voices are established in the first place, focused on how everyday lived experience intersects with power or social struggle. The "male sort of loneliness" associated with computing is in the end a product of history and context, not biology. Forms of selfhood bear the historical markers of their times. On the one hand, this means one needs to carefully explore the cultural patterns of meaning inherited by a given community. Most of the people who played key roles in developing the internet, for example, inherited a tradition in which technological mastery was imagined as inherently masculine. From railroads to radio, from automobiles to VCRs, mastery of technologies has been treated as a sign of male prowess and control. This history weighs on the cultures of engineering and policymaking discussed in this book and will be discussed at relevant points.

Yet cultural studies also effectively brought to the discussions of identity and difference a concern with the *relational* as opposed to essential quality of subjectivities. E. P. Thompson argued that social class—the social identity with which cultural studies began—needs to be understood as "a relationship, not a thing,"[29] "an active process which owes as much to agency as to conditioning . . . something which in fact happens."[30] So the focus of this book is less with what category people belong to in the fixed sense, less with their socioeconomic or ethnic backgrounds, and more with the dynamics of various constructions of selfhood in specific contexts. Cyberpundit Esther Dyson is a woman, but what mattered most about her activities in the 1990s was her libertarianism; she was a key figure in promoting and making acceptable the tendency to imagine society online as made up of abstract individuals pursuing their interests in a marketplace, individuals imagined as if their ethnic, gender, or class status did not matter. That libertarian model of selfhood—its allures and its limitations—played a key role in the trajectory of the internet and its reception in the early 1990s.

Of course, the libertarian's notion of individuality is proudly abstracted from history, from social differences, and from bodies; all that is supposed not to matter. Both the utilitarian and romantic individualist forms of selfhood rely on creation-from-nowhere assumptions, from structures of understanding that are systematically blind to the collective and historical conditions underlying new ideas, new technologies, and new wealth. For historical and sociological reasons, these blinkered structures of understanding have come more easily to men than to women. And it is in varieties of this gesture, I have found, that the forms of identity promoted by the cultures of the internet have most obviously played a role in the power dynamics of U.S. society. Various experiences of computing, from surfing the web from a cubicle to investing in the wildly expanding stock market, when coupled to various political discourses, have occasioned a revivification of an enthusiasm for the idea of the abstracted individual in the culture and a concomitant insensitivity to social relations and inequalities. The fact that computing can seem thrilling, that it can feel like an escape and thus like a type of freedom happens more often to men than women in the United States. But it is in that *process* of constructing what is experienced as seperateness, in the promotion and reinforcement of, say, "a male sort of loneliness," more than the simple statistical fact of male dominance in computing, that the flux of identities associated with the internet has mattered.

Ullman's depiction of a male loneliness associated with programming counters the individualist rendition of the experience of computing, not by denunciation, but by astute observation. Her writings are full of people alone at their computers who feverishly reach out over the wires for expression, connection, affirmation while ignoring the people around them—in the next room, across the street, in their city. Ullman's work in general, fictional and non-, uses a novelist's attention to human detail to tease out the inner fabrics of the experience of computing, those patterns of grandiosity, obsession, and discovery intertwined with moments of missed human connection characteristic of the last three decades of the computing culture in the United States.[31]

That male sort of loneliness, then, is of a piece with the fact that our culture has imagined personal autonomy to various degrees in terms of the model of the historic power of men over women, in terms of the power to command, to walk out the door, to deny the work of nurturing and the material fact of interdependence. It is this habit of understanding freedom negatively, blindly, as freedom from government, freedom from dependency, freedom from others, that helps set the conditions for the popularity of the rights-based free market. But, this book suggests, the same structures of self-understanding also set the conditions for the constructed sense of a lack, a felt absence, that can turn into romantic longing for some unknown or unachievable other. In that longing may lie seeds to change.

The Chapters

Many of the collective technological decisions that have constructed the internet have been gradual and are still underway. The discussion of them, therefore, is in various ways woven throughout the book. But each chapter centers on a particular set of choices and associated visions. Shared visions tend to evolve gradually between communities, without sharp boundaries in either time or space, so while the chapters are organized roughly chronologically, there is some overlap, and some simultaneous events appear in different chapters.

Chapter 1 sets the stage for the rest of the book by introducing several key concepts while exploring the early cultural and institutional contexts that set in motion the research and institutional support leading to the creation of the internet. Starting around 1960, it looks at the beginnings of the shift from the original vision of computers as calculating machines, in a category with slide rules, towards the idea that they might be communication devices, in a category with books, writing, and the telegraph. This conceptual change was crucial to the shift from centralized, batch-processed computers to the interactive, decentralized computers of today. At stake in these differences are competing visions of the character of human reason, particularly the problem of relating means to ends.

Are computers strictly a means to an end, or can they be an end in themselves, for example, a form of play? Early computers, this chapter shows, embodied and foregrounded this question for their designers.

Chapter 2 looks at how the initial discoveries of the playful possibilities of computing were seized upon in the late 1960s and early 1970s. In the wake of the 1960s counterculture, approaches to computing that loosened the connection between means and ends—that allowed play—helped create a subculture within the community of computer engineers. This in turn helped set the conditions for the rise of the modern, internet-connected, graphically-capable computer. The chapter introduces the theme of romantic individualism, an enduring Western cultural discourse with an associated way of imagining the self that passed from milieus like the San Francisco based counterculture, particularly that surrounding Stewart Brand and the *Whole Earth Catalog*, into the computer counterculture, as exemplified in the work of Ted Nelson, the computer visionary who coined the word *hypertext*. Against a background of Vietnam War era social disaffection, key romantic tropes—the strategic use of colloquial language, a studied informality, appeals to self-transformation instead of need-satisfaction, tales of sensitive rebel heroes, and a full-throated departure from instrumental rationality—became associated with alternative uses of computing.

Chapters 3 and 4 both focus on the 1980s. Chapter 3 considers events that happened very much in the public eye: the rise of the microcomputer envisioned as an icon of neoliberal marketplace enthusiasms, which helped justify the radical market-oriented policies of the Reagan era. The microcomputer revolution for the first time brought large numbers of Americans into direct contact with interactive computers, an experience framed by the historical accident of the stand-alone technical design of the machines and the entrepreneurial character of many of the businesses involved. Networking was ignored in part because the dominant culture was seeing things through free-market lenses and thus imagined that microcomputers were about isolated individuals buying and selling objects; this obscured the broader social relations like networking that both produced microcomputers and that could be enabled by them. The experience of first-generation microcomputers as distinct commodities thus helped articulate in the popular imagination a new sense of how a market vision reminiscent of the seventeenth-century philosophies of John Locke might be relevant to a modern, high-tech world.

Chapter 4 focuses on events of the 1980s that happened almost invisibly for most Americans: the development of an unusual culture of informal, open, horizontal cooperation—that very distinct set of practices that are incompletely summarized today under phrases such as "rough consensus and running code" and "end-to-end design." This chapter looks at two historically consequential but not often noticed instances of this set of practices. First, it looks at the development of new chip design methods in the late 1970s which led to VLSI (Very Large Scale Integration) microprocessors in the 1980s—the platform upon which the computing industry has grown ever since. The VLSI chip design process illustrates the discovery inside computer engineering of the sheer technical value of attention to social process—what engineers at the time called learning to "design the design process"—and the value of networked horizontal relations towards that end. Second, the chapter discusses the remarkable process by which ARPANET efforts were split off from the military and quietly transferred to National Science Foundation (NSF) funding. Theoretically, packet-switched global computer networking could have come to us in any variety of institutional packages, but this 1980s experience of quietly guiding the growing internet into a space between the differentially charged force fields of military, corporate, university, and NSF funding left a stamp on the institutions of the internet that would have far-reaching consequences.

Chapter 5 looks at the structure of feeling created in the early 1990s as knowledge workers began to discover the pleasures of online communication in substantial numbers, and elites groped for an organizational framework under the umbrella of the "information superhighway." Web browsing articulated itself with a structure of desire centered around an endless "what's next?" and spread in a context in which middle ranks knew things that their superiors did not, adding to that articulation a romantic sense of rebellion; one could in theory rebel, express oneself, and get rich all at once. Taken together, this fusion of romantic subjectivity and market enthusiasms, exemplified and enabled in the early *Wired* magazine, created the conditions that fueled both the rapid triumph of the internet as the network of networks and the dotcom stock bubble.

Chapter 6 looks at the rise to legitimacy of open source software production in the late 1990s. The open source software movement represents a rather sudden and dramatic transformation of dominant managerial principles in the high-tech industries. By 1998, Apple, IBM, Netscape, and others were investing heavily in open source software projects, actions that only a year or two earlier would have been considered laughably irrational. While there were economic conditions behind this, principally the Microsoft monopoly, those conditions also existed in 1996; economic forces alone cannot explain why the shift happened when it did. This chapter shows how the shift was enabled by a rearticulation of the romantic construction of computing through a retelling of the story of computer-programming-as-art that situated the narrative against, rather than for, the commodification of code. The effect of Eric Raymond's "Cathedral and Bazaar" essay and the spread of the rhetoric of open source associated with the Open Source Initiative were conditioned upon a widely experienced tension between the experiences of creating software and using computers and the structures of reward and industrial organization that emerged from commodified software; the same romanticism that had fueled free market visions earlier in the decade was now marshaled against them.

The conclusion summarizes the larger point to be learned from the previous chapters: our embrace, use, and continued development of the internet has been shaped by our experience of how it emerged. The openness of the internet is a product of the peculiar way in which it developed, not something inherent in the technology; the internet's history, as a result, is inscribed in its practical character and use. The internet has served as a socially evocative object for millions and created a context in which an ongoing exploration of the meaning of core principles like rights, property, freedom, capitalism, and the social have been made vivid and debated in ways that go well beyond the usual elite modes of discussion. It has played a key role in casting into doubt the certainties of both market policies and corporate liberal ones and widened the range of possibility for democratic debate and action, bringing to the surface political issues that have been dormant since the Progressive Era in the United States. But this efflorescence of openness is not the result of underlying truths about technology (or about progress or humanity) breaking through the crusts of tradition and inequality. It is the result of peculiarities of history and culture. The role of romanticism in particular reveals, not a universal truth, but the historical contingencies at work in the creation of both technology and democracy. As a practical matter, a new politics of internet policy making in the United States would be wise to take that history into account and start from the widely felt tensions between romantic and utilitarian individualism and move towards a richer, more mature approach towards democratic decision making.

1 "Self-Motivating Exhilaration"

On the Cultural Sources of Computer Communication

There is a feeling of renewed hope in the air that the public interest will find a way of dominating the decision processes that shape the future.... It is a feeling one experiences at the console. The information revolution is bringing with it a key that may open the door to a new era of involvement and participation. The key is the self-motivating exhilaration that accompanies truly effective interaction with information and knowledge through a good console connected through a good network to a good computer.

—internet pioneer Joseph Licklider[1]

Introduction

In a now-legendary moment in the history of the internet, when ARPA (Advanced Research Projects Agency) Director Jack P. Ruina was searching for someone to oversee the Department of Defense's research efforts in computing in 1962, he turned to J. C. R. Licklider. Ruina said Licklider "used to tell me how he liked to spend a lot of time at a computer console.... He said he would get hung up on it and become sort of addicted."[2] Today, in the early twenty-first century, that peculiar feeling of interacting with computers is familiar to millions; whether through computer games, web surfing, or actual programming, people across the globe have experienced versions of the compulsive absorption, the "addiction," that can sometimes come with computing. Hit some keys, get a response, hit again, another response, again, again, again. The little responses the computer offers—some numbers, an error message, an image, a sound—do not resolve things. Rather, they are just enough to invite the user to try again, ever in hope, in anticipation of getting it right, of finding what's next.

In the early 1960s, however, only a handful of people in the world had actually had the experience of interacting with a computer; computers were few, and only a small fraction of them allowed direct interaction through a keyboard and screen. Much of Licklider's uniqueness came from the fact that, among the very few who had directly experienced an interactive computer, he saw this "holding power"[3] as a potentially positive force; he eventually called it "the self-motivating exhilaration that accompanies truly effective interaction with information through a good console and a good network computer"[4] and famously proposed that it would eventually lead to more effective and rational human behaviors. As head of ARPA's computing project for most of the 1960s, Licklider went on to fund much of the research that laid the foundation for the internet.

Felt experience, as a rule, is generated by the interaction of embodied phenomena with cultural contexts. Ingesting peyote buttons, for example, can be experienced by American college students as a form of entertainment and by Southwestern Native American shamans as a connection with the ancestors. A mountain climber may experience the physical exhaustion of an arduous and life-threatening climb as supremely exhilarating; a war refugee fleeing her homeland might experience physiologically similar stress as a feeling of the darkest despair. The same is true of the compulsive experience of interacting with a computer. Computers exist in cultural contexts, in the equivalents of shamanistic rituals or college dorm rooms, and people give meaning to the technology accordingly.

This is not to say that computers do not have material effects apart from what the culture believes about them. In fact, it is probably the case that, economically and socially, the computers with the most impact on our lives are the ones that hum away quietly out of sight in large institutions, calculating our bank statements, connecting our phone calls, maintaining our payrolls, and the like. If those computers suddenly disappeared, our economic world would collapse. But if the computers on our desks, the ones we directly interact with and talk so much about, suddenly disappeared, for many of us it arguably would be little more than an inconvenience to go back to using telephones, typewriters, file cabinets, and photocopiers.[5]

For a technology to be integrated into society, however, *especially* when much of its activity is invisible, it has to be given meanings that can relate it to dominant social values, to everyday life, and to bodily experience.[6] Those often intricate meanings, in turn, shape over time what the technology becomes. Technology, in other words, is necessarily mediated by culturally embedded human experience. The meanings associated with the strange draw of the interactive computer are fluid. The fact that some see it is as an addiction while others see it as a potential source of human liberation, we shall see, is just the tip of the iceberg in terms of the various readings of the feel of interactivity. But the fluidity of these meanings is not random. They emerge out of particular historical and social contexts.

This chapter, while exploring some key moments in the early history of computing in the United States, introduces a number of ways of making sense of those contexts out of which the meanings of computer communication have emerged. It looks at the broad "corporate liberal" context of technological development in the United States, involving a pattern of quiet cooperation between corporations and government around technological issues, and traces the habit of instrumental rationality as a tool for organizing that cooperation. It then surveys the emerging tension between that instrumental rationality and the then-new experience of the compulsive pleasures of interactive computing, in which ends and means collapsed—which became expressed as a tension between the notion that computers were for addressing strictly defined problems consistent with established institutional goals and the dawning realization that they might be approached playfully, in a way in which the means *was* the end.

A Triumph of Personal Freedom or Strangelovian Nightmare? Problems in the Historiography of the Early Internet

One cannot say much about the past intelligibly without telling stories, without putting the past into narratives that give form to an otherwise inchoate mass of details. The stories told by historians, moreover, are rarely if ever incontrovertible; arguments continue over which story of the French Revolution best fits the facts. Yet, in the case of internet history, the stories told so far are marked by a peculiar tension; as historian Roy Rosenzweig has remarked, the stories people tell about the history of the internet seem torn between those that focus on its militaristic origins in cold war institutions and funding and the influences of the 1960s counterculture.[7] Take, for example, the story of Licklider. Science writer M. Mitchell Waldrop, in a 2001 biography, describes Licklider in warmhearted terms. "Lick," as Waldrop is fond of calling him, is presented as an amiable humanist and foresighted engineer and leader, a father of the internet who "provided the road map" for "the revolution that made computing personal."[8] As director of ARPA's computing efforts, Licklider was the man who "essentially laid out the vision and the agenda that would animate U.S. computer research for most of the next quarter century, and arguably down to the present day."[9]

Now take the version of Licklider in science historian Paul Edwards's 1997 *The Closed World*. According to Edwards, Licklider was significant principally because, as a key player in the military use and funding of computer research, he crystallized "cyborg discourse," a mode of thought born of the cold war that melded early computer technologies with "fictions, fantasies, and ideologies" of computer networks and artificial intelligence.[10] In Edwards's telling, Licklider is not Waldrop's amiable visionary that humanized computers but a cold war warrior committed to a system devoted to a dangerous dehumanizing dream of centralized military control of people and complexity via computerized control systems—principally for the purpose of fighting and winning a nuclear war. Edwards points out that most of the key innovations that spawned the internet—multitasking, time-sharing, interconnection of computers through networks, and accessible, interactive, graphically organized user interfaces through screens—came into being only because they lent themselves to a military vision of command and control. Edwards quotes a speech by General Westmoreland in the aftermath of Viet Nam. "I see," Westmoreland proclaimed, "an Army built into and around an integrated area control system that exploits the advanced technology of communications, sensors, fire direction, and the required automatic data processing—a system that is sensitive to the dynamics of the ever-changing battlefield—a system that materially assists the tactical commander in making sound and timely decisions." Edwards continues,

> This is the language of vision and technological utopia, not practical necessity. It represents a dream of victory that is bloodless for the victor, of battle by remote control, of speed approaching the instantaneous, and of certainty in decision making and command. It is a vision of a closed world, a chaotic and dangerous space rendered orderly and controllable by the powers of rationality and technology.[11]

This kind of Strangelovian vision energized by cold war urgencies, Edwards emphasizes, produced the vast and free-flowing sums of money that paid for much of the important research into computers as communication devices in the United States until the early 1970s and has remained an important impetus for research funding to the present day, particularly after it was reenergized by President Ronald Reagan's "Star Wars" project.

Here we have one of the major peculiarities surrounding the early history of the internet. One way of telling the story makes it seem to be about the gradual triumph of decentralized personal computing over centralized, impersonal computing; it is a story about the triumph of personal freedom, individual control, and uniqueness. Edwards's version finds the opposite: a story about visions of nuclear war and efforts to erase individuality through centralized command and control.

The difference here is not about facts; both authors do a solid job of getting their facts straight. Licklider did anticipate some features of present-day computing, and he also lived the most important moments of his professional career unapologetically near the center of efforts to build computer systems intended to fight nuclear wars. The difference is in how the story is told, in how the facts are woven together into a narrative. This tension or discrepancy pervades narratives about the internet; it calls for some serious reflection.

The Puzzle of Technological Innovation

Continuous technological innovation seems to be a key feature of the industrial world. The question of how technological innovations come about in general is no small one, nor is it purely intellectual. Technological innovation is a key measure of modernization, and as such its achievement is an ambition of practically every national government and most major political theories, from Marxism to

Reaganism. If one dismisses the simplistic extremes—for example, pure libertarian markets or pure centralized planning—the field of potential models of technology development is still quite diverse, complex, and interlaced with the passions of political visions. And in the history of technological change, the internet is a crucial case study, not the least because it has called into question many of the earlier received models of innovation. Like the unanticipated collapse of the Soviet Union, the surprising success of the internet has forced us to rethink what we thought we knew about how big changes happen.

It remains popular, certainly in the press, to recount histories of inventions by telling the story in terms of heroic individuals who, through sheer force of determination and genius, were right when everyone else got it wrong. School children are regularly treated to the story of the Wright brothers developing the first airplane in a bike shop, Edison inventing the light bulb, or Marconi the radio. This inclination also sometimes leads to a listing of "firsts": Westinghouse's KDKA was the "first" broadcast radio station, Paul Baran invented packet switching for postnuclear communications in the early 1960s, and Marc Andreessen created the first graphical web browser, Mosaic, in 1993 at the University of Illinois. Indeed, internet history is full of appealing stories of clever underdogs struggling against and eventually out-smarting various hulking bureaucratic military and corporate monoliths. Licklider is intriguing enough in his own right, but there is also the sincere political idealism of Douglas Engelbart, credited with the mouse, windowing interfaces, and other now-familiar features of computers. There's the remarkable collaborative spirit of the early ARPANET pioneers, who created the tradition of the gently named RFCs—Request for Comments—which set an enduring, casually democratic tone for developing internet technical standards. There's the joyful smart-aleck counterculturalism of Ted Nelson, who coined the term *hypertext* and who played a key role in popularizing concepts that are now embedded everywhere in microcomputers and the world wide web.

Yet, historians of science and technology have taught us that this way of recounting technological history has severe limitations. There are geniuses, no doubt, but the story is generally more complicated. As Robert Merton pointed out in his essay on "singletons and multiples," many innovations seem to occur to several different people in the same period.[12] Marconi was just one of numerous individuals who were developing radios at the turn of the twentieth century, and Brian Davies independently published a proposal for packet switching at roughly the same time as Baran. The Wright brothers may well have created the first working airplane, but if they hadn't, someone else eventually would have.[13] There were other light bulbs before Edison's, it turns out, other radios before Marconi's, other broadcast stations before KDKA, other proposals for packet switching, and other web browsers that predated Mosaic.[14] A typical search for the first instance of a particular technology tends to uncover a gradual evolution instead of a sudden sharp appearance and groups of individuals working simultaneously rather than unique individuals working entirely alone. Individuals do make unique and crucial contributions, but they generally do so only in the context of groups, of communities of interacting individuals working on similar problems with similar background knowledge. Innovation, then, is a social process. Seeking to account for these patterns, historians of science and technology have pointed to what they call "invisible colleges," dispersed communities of innovators sharing knowledge through journals, professional associations, and other forms of contact.[15]

The observation that invention is social, however, raises as many questions as it answers. Knowledge and new ideas flow both up and down the chain of command. For every famous leader or key publication, there are always many unknown or barely known individuals who are quietly working away in the background, often actually constructing machines and exploring new uses in ways that leadership comes to understand only after the fact. The degree to which Licklider was a brilliant creator of new ideas or merely a conduit for the ideas of others is difficult, perhaps impossible, to determine with absolute certainty. Furthermore, "society" is no more a full explanation of invention than is individual genius. The internet and computer networking may have originated in a military context, for example, but that does not mean those original intentions are embedded in their character or social function.

There is a danger, furthermore, in interpreting things teleologically, that is, in looking through the details of the past for events that seem to point to the way things are in our own time, as if our current situation was foreordained. One can find in Licklider's early works mentions of things that remind us of what has happened since. By filtering the details of the past through the lenses of the present state of affairs, however, an overly clean picture is painted of a linear trajectory leading inevitably towards our current condition. As we will see, Licklider had some ideas that were prescient, but others that seem odd, and his predictions were certainly far short of exact, and in any case his principal impact came through teaching and the granting of funds, not the actual building of equipment. Teleological accounts assume away the possibility that things are the way they are to some degree because of an accumulation of accidents, that current conditions need not be the product of some ineluctable law of history, that they are contingent and could be different. For this reason, it is important also to look beyond the moments that in retrospect appear prescient, to roads *not* traveled and to collective mistakes.[16] History needs to look, not just at what people got right, but what they got wrong or merely differently. There are numerous stories of technologies developed for one purpose that ended up being used for another: the telegraph and the airplane were expected to end strife between nations, Alexander Graham Bell imagined the telephone would be used for a kind of wire-based broadcasting, and early developers of radio transmission of sound imagined they were creating a wireless telephone. (At the same time, such changes in the imagined use of technologies need not be proof of technological autonomy, as though technological development occurred somehow outside social process and purposes.)

So the task is to specify the complex interactions between social context and technological potentials and developments, the layering of social vision with technological processes over time. Somehow, the evidence suggests, certain discoveries and innovations seem to be in the air at specific times and places, ready to emerge through the efforts of properly skilled and situated individuals. The effort to look at what's in the air, then, is an invitation to explore complexity, not to reduce things to a singular social cause. And this requires taking a step back to look at broad historical patterns of technological development.

Building Big Systems the American Way: Corporate Liberalism and the Military-Industrial-University Complex

Part of the standard folklore about the internet's origin is that it was invented as a form of communication that could withstand a massive nuclear strike because the system would keep functioning even if individual nodes were randomly eliminated. This is too simple. As some of the key individuals involved have pointed out,[17] the concepts and technologies at the internet's heart were actually developed with much broader applications than postnuclear survivability in mind. Yet this folklore persists, in part, because it is a convenient way to explain the "distributed" character of the internet's structure and, in part, because there's a partial truth to it: one of several sources of the concept of computer-based packet-switched communication networks was a series of articles published by Paul Baran on "distributed networks" for the RAND corporation, which indeed were motivated by a perceived need for communication networks that could survive a nuclear attack.[18]

But, on another level, the persistent myth of postnuclear survivability as a source of the internet resonates with a deeper truth: the internet was to a large degree created by people within or funded by the U.S. Department of Defense's Advanced Research Projects Agency (ARPA, with a D for Defense for a time added to make it DARPA), a cold war institution created in the 1950s specifically to counter the Soviet Union's perceived technological superiority in the wake of Sputnik. Even if it wasn't invented only with postnuclear applications in mind, military interest in advanced technological research in the post-WWII era went far beyond specific applications. Rather, it was of a piece with a sociopolitical belief system that had nuclear weaponry at its heart.

In 1945, MIT scientist and administrator Vannevar Bush wrote a widely pub-licized report about a postwar policy for scientific research.[19] Bush's *Science, the Endless Frontier* drew on his experience of World War II, particularly organizing the Manhattan Project. Bush, a political conservative, was no Einstein or Oppen-heimer, who agonized over the moral consequences of the construction of the atom bomb. To Bush, the Manhattan Project was simply a rousing success and could serve as a model for further technological development of all kinds. The U.S. gov-ernment, Bush argued, should be in the business of funding basic research in sci-ence and technology. Private enterprise, the argument went, would be unable to take on the risks of basic exploration because it was too uncertain to justify investment. Government and nonprofit institutions like universities and the military, therefore, should conduct the initial, high-risk exploratory research and then turn the results over to industry to develop commercially exploitable applications; government-sponsored research yields practical benefits that can eventually be exploited by the business world. As Bush put it, "There must be a stream of new scientific knowledge to turn the wheels of private and public enterprise."[20] Bush's argument proved influ-ential, helped inspire the creation of the National Science Foundation and similar institutions, and shaped the broad thinking about technological innovation in the United States for the last half-century. And the policy Bush outlined has been suc-cessful; the pattern of developing technology with initial public money followed by commercialization has brought us satellite communication, microwave ovens, computers, jet airplanes—and, to a large degree, the internet.[21]

The Bush approach to technology development was hugely influential, but it was not without precedent. It is symptomatic of a general trend in twentieth-century American political economic thought usefully called corporate liberal-ism[22] and is associated with the broad idea that government and industry should cooperate at key moments in the name of furthering the intertwined processes of economic and technological development. The idea underlay Herbert Hoover's efforts as Secretary of Commerce in the 1920s to use gentle government regula-tion to foster the development of a commercial radio industry and, in a rather different way, Franklin Roosevelt's efforts to prime the economic pump through government funded technology projects like the Tennessee Valley Authority.

Corporate liberalism is corporate both because it coincides with the rise of great corporations as the dominant economic organizational form and because it involves a certain collective vision and coordination across public and private institutions. It is liberal because it seeks to square that collective vision with tra-ditional individualist, liberal principles, like free enterprise. Corporate liberalism, so defined, goes back to the turn of the previous century. Its existence is confirmed by the historical interdependence of corporations and government, which is why the growth of big business in the twentieth century closely parallels the growth of big government. It is not a tightly organized, entirely rigid system, but more a set of habits cultivated over decades (often in a sort of teeth-gritting harmony with interest groups representing various forms of underprivileged sectors of society who regularly use the language of corporate liberalism, such as "the public interest," to advance their own goals).[23]

These long-standing patterns of business-government cooperation are rarely acknowledged in broad public discourse.[24] Yet there exist communities comprised of key individuals in management, engineering, and government for whom it is hardly a revelation to point to sustained business-government cooperation. Behind all the storm and fury that rages between advocates of "markets" versus "government" approaches to industry, key communities of corporate and public leadership have quietly gotten used to the notion that private and public institutions can and should cooperate effectively in certain areas. Long-standing communities of practice have developed that take a certain amount of cooperation for granted and find the ambiguities of the zone where private and public sectors meet to be productive.

One of the key areas for this business-government cooperation has been that zone of high-technology development encouraged by Vannevar Bush. In a culture where both the Left and the Right tend to imagine the public and private sectors as radically opposed to one another, it can be quite illuminating to hear engineers talk thoughtfully about the relative strengths and weaknesses of working for businesses, universities, and government agencies. Participants in the early development of the ARPANET, for example, speak fondly of the "loose" oversight and exploratory approach that characterized both the public sector and private sector managerial framework; key individuals at ARPA felt free to ignore requests from military officers that seemed foolish, and on occasion they would ignore official titles and job descriptions for the sake of working with individuals viewed as interesting and adept.[25] Bolt, Beranek, and Newman, the private consulting firm that was the principal private contractor for the ARPANET, was known by some insiders as the third university in Cambridge, Massachusetts (after Harvard and MIT) because of the academic atmosphere.[26] Historian of technology Thomas Hughes speaks with admiration of how in the post–World War II era "the conviction that they were responding to a national emergency" of the cold war allowed pre-1970s developers of advanced military technology to "energetically and effectively [counter] the bureaucratic tendencies common to large projects and organizations."[27]

The Vannevar Bush approach to technology development is a reigning idea, however, not a popular one.[28] Using large influxes of government funding to develop the basics of technologies for later exploitation by industries makes good sense to acronym-fluent professionals who work inside large organizations like universities, high-tech corporations, the National Science Foundation, and the Pentagon. Yet it is the main rationale for the military-industrial complex, and that term remains an epithet across the political spectrum. Bush's approach to technology development flies in the face of some core American values.

Basically, the Bush approach willfully blurs boundaries between public and private sectors and turns decision making over to small clubs of insider firms and institutions. At odd but inevitable intervals, as a result, it comes under fire from both the Left and the Right. Republican President Dwight Eisenhower's 1959 speech in which he coined the term "military-industrial complex" as a warning was just one early moment. In the late 1960s, the new Left made criticisms of the military-industrial complex a regular part of its repertoire. And by the 1980s right-wing antigovernment proclivities in the Reagan administration led to reductions in federal funding for basic research coupled to calls to rely on the private sector alone.

In recent years, Thomas Hughes has been a somewhat lonely voice speaking explicitly on behalf of what he calls the "military-industrial-university complex," the nexus of high-technology corporations, university research programs, and government agencies that together have shepherded in many of the great technological achievements of the last half-century.[29] And he has spoken with some alarm and dismay at the post-Vietnam generation's general disparagement of this complex and at the resulting two-decades-long decline in federally funded research in the United States. He persuasively argues that much of the animus against government funding for research is based on a misunderstanding of the nature of technological innovation. The astonishing advances in microchip design and manufacture as well as telecommunications of the last thirty years, he points out, are not products of the market acting alone; in each case, key advances were incubated by government funding.

Under the auspices of the National Research Council, chartered to advise the government on matters of scientific and technological research, Hughes chaired a committee that prepared a 1999 report *Funding a Revolution: Government Support for Computing Research*.[30] In an effort to refute the neoliberal distrust of government and the associated belief that the market is the cause of all innovation, the report exhaustively details the fundamental role of federal funding and institutions in creating the computer revolution, from networking technology to VLSI microprocessors to computer graphics. The report is incontrovertible in its general conclusions that initial government money was essential to most major developments in computing. In sum, from the point of view of certain managers and engineers, the argument goes, the blurring of boundaries between public and private inherent in the Vannevar Bush approach can be experienced as a strength. The very ambiguities that create so much trouble for the American political imagination, in sum, also seem to create a moderately "open" space for scientific and technological inquiry.

But Hughes' effort to find wisdom in the military-industrial complex pits him against a trend that cuts deep into the American psyche: a profound American ambivalence about government and big organizations in general. Consider the fact that, although Edwards and Waldrop provide polar opposite accounts of Licklider, both narratives are animated by an aversion to big government-funded, bureaucratic organizational structures. In Waldrop's story, Licklider and his fellow enthusiasts of "personal" computing are creative individuals who triumphed through pluck, brilliance, and a resistance to blinkered bureaucratic formalities and political concerns; in Edwards's story, Licklider embodies that blinkered bureaucracy. But neither writer shows signs of wanting to stand alongside Hughes in the position of more or less *approving* of that bureaucracy.

American ambivalence towards corporate liberal patterns need not be dismissed as simplistic. Is corporate liberalism good or bad? is a good question; it is not irrelevant that many of the technological accomplishments that Hughes celebrates did not work (for example, SAGE) or might have ended human life as we know it if it did (the intercontinental ballistic missile). But it is a complicated question, and the beginning of an answer is only going to be found by going beyond the various competing conventional wisdoms on the matter and looking carefully at its operations in detail. In some sense, this book views the internet precisely as a case study of corporate liberalism in action. Something was done right in the early days of the internet's development; the internet is a case of "good government," of the effective use of certain aspects of government power to improve the lot of humankind. But what happened does not follow the vision of Vannevar Bush exactly; a bit of reflection is required to understand what exactly was done right.

The Culture of the Interface: Batch Processing versus Interaction

Edwards's *Closed World* still stands almost alone in emphasizing, not just the role of military funding, but the role of military designs and intentions, in setting the stage for the rise of modern computing. Edwards points to the darkest side of the military-industrial complex, a side that most discussions of early computing are too quick to pass over lightly, if they mention it at all. But Edwards's book was largely written before the internet became what it is today, so he does not directly address what is now an obvious question: How could the "closed world" have led to the creation of the "anarchistic" internet? Part of what's missing from Edwards's account is a sense of what students of culture call the lived experience of people working on computing: the feel of working with computers and the meanings that various communities attached to them.

Systems "Science" and Data Processing

The common way of using computers in the 1950s and 1960s was batch processing. Typically, a user would prepare a stack of punch cards, each containing a separate instruction or field of data, give those to a computer operator, and hours or days later pick up a printout with the results. This practice partly reflected the fact that computers were hugely expensive, delicate, and limited, so that allowing anyone besides specially trained operators direct access to the machines would have been unacceptable. Batch processing was a way of rationing access. Varieties of batch processing persisted well into the 1970s, even as interactive terminals started to become available. (Users of MS-DOS may remember .bat files that could contain a series of instructions to be executed in sequence; the .bat stood for *batch*.) But batch processing, intentionally or not, also lent itself to certain habits of thought. It fit neatly with the original notion that computers were for doing elaborate calculations, in the way a scientist or statistician would develop a formula or field of data before running the calculation itself.

The 1960s was the period in which computing spread into the corporate world, largely in the form of expensive mainframes. Banks and other large enterprises discovered that computers were useful for maintaining and manipulating forms of information, such as mailing lists of customers and records of financial transactions. In the 1960s these were the fastest growing areas of the industry. Spearheaded by IBM, computing became a part of big business and also became a big business itself.

The model of batch processing held sway during this era, partly for the practical reason that it rationed computer use but also because it fit the institutional ethos of large corporations. The tasks the computers were being used for—coordination and rationalization of giant vertically integrated corporate enterprises—called for uniformity and predictability across large organizations and thus tightly delineated, prespecified tasks and formats. The separation of means and ends is a hallmark of what Habermas and others have described as "instrumental rationality," the logic that has been associated with bureaucracy and large hierarchical organizations; goals and procedures are worked out beforehand, and then everyone and everything just implements them according to narrow rules. Large machines in the basements of corporate headquarters seemed to embody this logic. They predictably carried out routine tasks that were tightly specified beforehand from on high; this fit the corporate model to a T. These computers were not celebrated as fun; they were imagined as powerful. The general sense of computing at the time reached its fullest pop-cultural expression in HAL, the murderously intelligent computer in Stanley Kubrick's film *2001*.

One of the ironies during this period, however, was that as computers became more common they were doing less actual computing, less mathematical calculation. Instead they were sorting, organizing, and comparing—doing those things we now associate with databases. Increasingly, the underlying logical structure was not mathematics but sequences of letters and words, particularly the arbitrary sequence of the alphabet. By about 1970, the majority of computers were used more for manipulating symbols than they were used for complex calculations, less for number crunching than for what was coming to be called data processing. Licklider and his associates would occasionally use this trend to make the case for thinking of computing in radically new ways, more as communication devices than as fancy calculating machines.[31] But this did not fit most people's preconceptions. People still tended to assume that computers were fundamentally mathematical tools. Computer science professor and pioneer Andries Van Dam once approached his university's vice president in charge of computing in the 1960s with a request for computer time for experiments in using computers for the humanities. The administrator was reluctant because "that would subvert the true purpose of computers, which was to produce numbers for engineers and scientists." "If you want to screw around with text," the administrator continued, "use a typewriter."[32]

What sustained this vision was that, instead of asserting that computers were *not* strictly mathematical, the logic tended to work in reverse; computers, it was assumed, were bringing an aura of mathematical certainty into nonmathematical problem areas. Computers would allow us to "mathematize" human affairs. So the nonnumerical use of computers fell under the stilted rubrics of data processing and information processing. Data processing, it was often implied, would bring mathlike scientific precision, efficiency, and control to ever more areas in life.

This dream was near the heart of the grandest and most influential intellectual framework for making sense of all this at the time, systems science, or cybernetics, which reached its apogee in the late 1960s. The core conceit of systems science was that nearly everything, from ballistic missiles to corporate hierarchies to political processes, could be conceived and quantitatively analyzed in terms of functionally organized systems of feedback loops. The use of computers for managerial and military command-and-control and their use for more mundane tasks like maintaining an inventory were understood, in sum, as simply variations on the same theme. Much of the thrill of systems science came, not just from its promise of understanding complexity, but from its promise of control of that complexity. The conceits of systems science allowed communications at a distance and control over human affairs to be understood as compatible, almost integral. Popular both in academia and in the world of think tanks and foundations like the RAND corporation, systems science was offered as the solution, not just to problems of military complexity, but also problems like inner-city strife, poverty, and inequality.[33]

But even without the grandiosity of an overarching framework like systems science the dream of mathematizing mundane aspects of life filtered into many areas of life. A small but telling illustration of how the vision embodied in information processing operated in practice can be seen in the early efforts around what was called office automation, a trend that appeared during the decline of batch processing but that pursued a similar underlying logic. In the late 1960s and early 1970s, the computer industry, with IBM at the lead, introduced the concept of centralized word processing centers, where traditional secretaries were supposed to be replaced by teams of "correspondence secretaries" working at rows of terminals attached to larger computers in the name of efficiency and cost reduction.[34] This vision became referred to as the office of the future, and other high-tech firms eventually sought to enter this apparently burgeoning market. Computers, the theory went, would automate the front office the same way machines and engineering had automated the factory. *Forbes* enthusiastically predicted, "with automatic typewriters rattling off error-free letters at incredible speeds, it will just take fewer secretaries to do the job."[35] In 1973, Xerox Chairman C. Peter McColough described the company's strategy this way: "In the next decade, if we are to generate real efficiencies in the office, we're going to have to alter traditional structures. The idea of one secretary for one executive is no longer efficient or economical. And we have to reduce and reposition the role of paper."[36]

Within a decade, McColough's vision would become a popular business school case study in managerial short-sightedness.

"Man-Computer Symbiosis" and the Limits of Instrumental Reason

Licklider was part of a minority with other ideas. The principle exception to batch processing at first came directly from the cold war military efforts of the 1950s. The military emphasis on jets, radars, and eventually ballistic missiles created a need for speed. By the early 1950s, an effort began to develop computers that could be used to calculate trajectories and flight paths of missiles and planes almost instantaneously so that operators could use the computers in much the same way radar technology was used in World War II, but across greater distances and with greater accuracy. The disciplinary term for this field of endeavor became *communications, command, and control*; Licklider's official title at ARPA was head of the Behavioral Sciences and Command and Control programs. The goal was to use computers and telecommunications technology to extend military control across distances with ever more detail and in ever quicker response times, bringing power towards the center and reducing the autonomy of those on the front lines. An early experimental computer built at Lincoln Laboratories at MIT called Whirlwind eventually grew into project SAGE, the first part of the nuclear early warning system.[37] SAGE was not a success in its own terms—many now say that the system never would have worked in the case of an actual nuclear exchange[38]—but the mode of using these computers uniquely involved a radar-like cathode ray tube or CRT. SAGE was what people now call interactive. Thus it was that the first people to experience "playing" with a computer did so in the heart of the cold war effort to manage the unmanageable possibility of nuclear war.

This was the technology with which Licklider first experienced the addictive quality of direct computer interaction. Some who noticed the holding power of direct interaction with the computer no doubt treated it as odd but insignificant, one of those things you might notice but then dismiss with a shrug while you go on to other matters. Licklider's intellectual uniqueness may have lain in his effort to attribute positive *meaning* to the experience, to put it in a specific, legitimating intellectual framework. There were others of his generation who made specific technological innovations that contributed to the new way of conceptualizing computers; Ivan Sutherland's work on computer visualization, for example, was central. But Licklider seems to have been the one who provided a framework that helped others to look at innovations like Sutherland's and see something other than a curious gizmo. Licklider was a leader in taking his fascination with the interactive experience and seeking to turn it in new, more generalized directions.

Joseph Licklider was trained as a psychologist in the 1930s and 1940s, with a particular interest in how people process information. One of the key moments in the development of the idea of systems science or cybernetics was a series of meetings called the Macy conferences held immediately after World War II in Cambridge, Massachusetts. Norbert Wiener, Licklider, and many other foundational figures attended these invitation-only meetings that gave birth to a highly influential set of terms and concepts that, as Edwards puts it, "redefined psychological and philosophical notions in the terminology of communications engineering."[39] Licklider and his ilk would take the notions that emerged in these meetings—negative feedback loops, inputs and outputs, and so on—into the "Closed World" intellectual universe that metaphorically melded humans with machines for military purposes. The idea that thinking was like information processing laid the foundation for both the idea that computers could become minds, the cornerstone of the field of artificial intelligence, but also, reciprocally, that people could be understood and organized into centrally controlled communication systems, the cornerstone of the field of systems science.

Licklider's principal influence came from his role as head of computing at ARPA in the 1960s, where he gave out a lot of money to people who would eventually develop many of the practices and protocols that became the internet. But indicative of his thinking, and of the rhetorical environment of the time, was a 1960 essay called "Man-Computer Symbiosis." This essay is said to have been a galvanizing synthesis of cutting-edge thinking about computer interactivity.[40]

Most read this essay today in classic teleological fashion, scanning the article for elements that seem to predict our present-day situation, such as Licklider's call for "Desk-Surface Display and Control"—something some have seen as a forerunner to desktop computers—or his proposals for graphics and icons to facilitate computer use. On this basis, the article is frequently cited as a key text in inspiring the modern computer revolution.[41] While it is certainly true that Licklider was looking for alternatives to batch processing earlier than most in the 1950s and early 1960s, one needs to look closely at why.

"Man-Computer Symbiosis" is a strange piece. It is certainly not a direct contribution to science or engineering; it contains only the most casual references to actual technological developments and capacities, and his primary empirical evidence is a so-called time and motion analysis that is little more than Licklider's description of how he spends his day. And the conceit of the title—that humans and computers could interact in a symbiotic relationship, as if they were both living beings—has rarely been taken seriously since.

Paul Edwards offers one of the only critical discussions of this essay, focusing on its cold war institutional context and its cyborglike vision of the relation between humans and computers, with its implied reduction of human nature to predictable and controllable systems.[42] The title's metaphor, Edwards notes, is basically a twist on a very science-fiction-like notion of artificial intelligence. Licklider casually cites a prediction that computers will surpass the power of the human brain by 1980—one in a long line of wildly overoptimistic predictions in the field—and suggests man-computer-symbiosis as a strategy for the interim period before computers reach that point.

Part of what makes the article work is Licklider's suggestion that direct interaction with computers could automate drudgery; instead of spending time plotting graphs by hand, for example, scientists and managers could let computers draw the graphs for them and spend their energies on interpreting results. But Licklider takes care to distinguish symbiosis from the notion of computers as merely extensions of human capacity. His idea is largely that computers will work as happy, conversational slaves, taking care of all the tedious routine work that normally fills the life of a knowledge worker so that people can focus on actual learning and decision making.

One of Licklider's two principal justifications for his scheme is drawn from cold war military designs that were paying his bills. "Imagine trying," he writes, "to direct a battle with the aid of a computer" using batch processing. "Obviously, the battle would be over before the second step in its planning was begun. To think in interaction with a computer in the same way that you think with a colleague whose competence supplements your own will require much tighter coupling between man and machine than is suggested by the example and than is possible today."[43]

But Licklider's other justification, which he lists *prior* to the military one, is this:

> Many problems that can be thought through in advance are very difficult to think through in advance. They would be easier to solve, and they could be solved faster, through an intuitively guided trial-and-error procedure in which the computer cooperated, turning up flaws in the reasoning or revealing unexpected turns in the solution. Other problems simply cannot be formulated without computing-machine aid. Poincaré anticipated the frustration of an important group of would-be computer users when he said, "The question is not, 'What is the answer?' The question is, 'What is the question?'" One of the main aims of man-computer symbiosis is to bring the computing machine effectively into the formulative parts of technical problems.[44]

This is a significant passage with an interesting philosophical twist. Licklider is trying to argue, in language amenable to the "Closed World" logics of his time and place, that unscripted *playing* with a computer might be useful. Play involves, and might be defined as, an activity engaged in for its own sake instead of for a prespecified end. Licklider wants computer systems that are interactive because they lend themselves to tinkering, to fiddling, systems where you can play with them without having a tightly specified plan or goal. Such a view of computing, whatever else one could say about it, would offer Licklider a *meaning* in his experience of computer compulsion. It would justify his pleasure in the machine.[45]

In retrospect, we could reasonably argue that much of "Man-Computer Symbiosis" is odd, wrong, or thinly supported. The article's influence, though, was not due to its specific technological arguments. His effort to articulate a justification for his pleasure in the machine brings him, and the reader along with him, to the edges of the dominant logic of his intellectual environment. Licklider is no philosopher, but by suggesting that question-formation might be brought into the computing process—the question is, what is the question?—he is expressing a certain wise frustration with the technocratic ideal of instrumental rationality characteristic of the institutional world he inhabited. Batch processing as a practice enforced a strict separation of means and ends; for any given problem, the questions and methods for answering them had to be worked out entirely before the stack of cards could be submitted to the computer. Licklider is looking to justify a computing practice that is not about achieving pregiven, strictly delineated ends; he is looking for an alternative to the narrow, instrumental rationality that was reinforced by traditional batch processing.

Licklider's desire to justify unscripted play with the computer took him to the edge of the dominant logic of his time, but not beyond it. The matter-of-fact references to fighting wars, to commanding and controlling distant events, are still there alongside the brief hints of alternative logics; it says something of Licklider's time and place that he can see these different modes as congruent (and something about ours that we cannot). By 1965, the concern for winning battles began to be sidelined in Licklider's writings, but there was still a general emphasis on an abstracted individual asserting control at a distance. He argued, for example, that a computerized information system would bring "the user of the fund of knowledge into something more nearly like an executive's or commander's position. . . . He will say what operations he wants performed upon what parts of the body of knowledge, he will see whether the results make sense, and then he will decide what to have done next."[46] The user, even in engaging in personal or political exploration, would be commanding that operations be performed and will have things done. The reasoning was instrumental, even if the applications were no longer so blatantly nightmarish.

By the mid-1960s, though, Licklider also began to add flights of utopianism to his descriptions of the possibilities of networked computing. He argued that, if a massive computer network involving "home computer consoles" and television sets were constructed, citizens would be "informed about and interested in, and involved in, the process of government. . . . The political process would essentially be a giant teleconference, and a campaign would be a months-long series of communications among candidates, propagandists, commentators, and voters."[47] What's significant is not just that Licklider anticipated the blogosphere forty years in advance—that would be simple teleological analysis—but that this brief sketch emerged as part of an effort to justify unscripted interaction with computers, an effort to overcome the separation of means from ends characteristic of instrumental reasoning.

Engelbart, Bush, and the Encyclopedic Dream

Here Licklider may have been echoing the thoughts of one of his protégés, Douglas Engelbart, who by that time was working from a lab at the Stanford Research Institute in Palo Alto. Engelbart is known today as the inventor of the windowing interface, the mouse, and the idea that networked computers could be used as collaboration devices. There is no question that his work was important and influential. But we should not fall prey to reading events through the lens of the current structure of things or conclude that his visions or specific inventions had impact simply through their brilliance or foresight. Engelbart and his team were certainly not the only ones working on computers as communication machines. For example, while Engelbart continued working on his project in the 1970s, computer scientists at the University of Illinois were developing the PLATO system, which was focused on educational uses and that used touch screens to navigate graphical interfaces; it may be merely accidents of timing, funding, and geography that we all use computer mice today instead of touch screens.[48]

Yet Engelbart's research is indicative of the way that dominant logics and lived experience of technologies interact. Engelbart's efforts, though working with ARPA funding funneled his way by Licklider, show little or no direct evidence of the cold war command and control vision evident in Licklider's writings. (Edwards does not discuss Engelbart.) Instead, Engelbart insisted that his work on a "Program on Human Effectiveness" was necessary to social improvement. Taking Licklider's interest in improving human problem-solving abilities to a grander level, Engelbart introduced his program in this way:

> Human beings face ever more complex and urgent problems, and their effectiveness in dealing with these problems is a matter that is critical to the stability and continued progress of society. A human is effective not just because he applies to a problem a high degree of native intelligence or physical strength ... but also because he makes use of efficient tools, methods, and strategies. These latter may be directly modified for increased effectiveness. A plan to systematically evolve such modifications has been developed at Stanford Research Institute. . . . The possibilities we are pursuing involve an integrated man-machine working relationship, where close, continuous interaction with a computer avails the human of radically changed information-handling and -portrayal skills, and where clever utilization of these skills provides radical changes in the way the human attacks problems. Our aim is to bring significant improvement to the real-life problem-solving effectiveness of individuals. It is felt that such a program competes in social significance with research toward harnessing thermonuclear power, exploring outer space, or conquering cancer, and that the potential payoffs warrant a concerted attack on the principal problem areas.[49]

While still making the case for a "close, continuous interaction with a computer," there was only a mild echo of Licklider's "symbiosis"—"man-machine working relationship." Engelbart eventually described his efforts grandly as a system for the "Augmentation of Human Intellect."

More clearly than Licklider's vision, Engelbart's project is at least in part an heir to the Enlightenment fascination that found its classic expression in Diderot's *Encyclopédie*. The eighteenth-century French *philosophes* who contributed to the *Encyclopédie* hoped that rationally organizing and making accessible all the scientific and technological knowledge of the day would overthrow superstition and irrational passion, empower individuals, and advance human progress. Bringing all modern knowledge together effectively, the theory went, will lead to more rational and effective behavior by individuals and thus a better society. Diderot's pioneering *Encyclopédie*, moreover, was not the "linear," single-author book stereotyped by today's postmodern fans of "nonlinear" new technology; it had over a hundred authors, copious charts and illustrations (the multimedia of the day), and was intended to be cross-referenced and consulted, not read cover to cover. Engelbart was an inheritor of this tradition and shared its faith that a rationally organized, accessible system of knowledge will tame the bewildering complexity of the world and thereby overcome human folly.

The encyclopedic dream has been an ongoing feature of Western thought since the *philosophes*. In the 1920s, some librarians had been touting the powers of the then-new technology of microfilm for information storage along encyclopedic lines; microfilm would make hordes of information widely accessible and easily retrievable, thereby spreading enlightenment. Perhaps picking up on this theme, in the 1930s Vannevar Bush had begun speculating about the advantages of microfilm, which would allow "the contents of a thousand volumes [to be] located in a couple of cubic feet of desk, so that by depressing a few keys one could have a given page instantly projected before him."[50] Bush, the technocratic manager of massive engineering projects, had observed the difficulty of making one's way through vast amounts of technical information. "The investigator is staggered by the findings and conclusions of thousands of other workers," Bush wrote, but "the means we use for threading through the consequent maze to the momentarily important item is the same as was used in the days of square-rigged ships."[51] And then, in a piece first drafted in 1939 and published in *The Atlantic Monthly* in 1945, Bush sketched out a fantasy office machine called the "memex" that would automate what one does in a library—look up information, peruse indexes, take notes, and so on—all at the push of a few buttons embedded in a desk.

The memex, as Bush described it, was not digital, not networked, and not even workable. Its resemblance to the modern personal computer is in most ways superficial. Thierry Bardini, in fact, has called into question the idea that it was an important factor in the development of modern personal computing.[52] Bush should perhaps get credit, however, for advancing the idea that a machine like the memex could have the capacity to construct "trails" between documents and other bits of information. This was likely the first mention of something like hyperlinks. The originality of the idea of a hyperlink should not be exaggerated. The idea of a trail or hyperlink is just a variation on the idea of the cross-reference, performing a similar function to the footnote, the file card, or the index. It is not the case that before hypertext all books were read linearly, from front cover to back, without any attention to interconnections. No one has ever read books containing legal documents from cover to cover, for example; legal information has long been organized in a way that allows one to follow a web of trails into other documents to find relevant precedents, arguments, and so forth. Bush simply added the thought that a machine might automate the process of cross-referencing and suggested that such a machine might put that power in the hands of readers as well as writers and editors. So to an important degree, Bush and Engelbart were continuing and extending the traditions of the Enlightenment encyclopedic ideal, not departing from it.

What Engelbart and Bush's memex proposal clearly share, however, is a belief that the key problem was not the murk of medieval superstition and traditions, the chief bugaboo of the *philosophes*. Rather, the problem was too much information, poorly organized. This was their new twist on the encyclopedic dream. The traditional apparatus of footnotes, libraries, indexes, and file cards, in other words, had not produced the world of enlightened clarity the *philosophes* had imagined. In a sense, the original encyclopedic project had been realized, but it did not work; the modern world was chock-full of, not just encyclopedias, but entire libraries bursting with indexed knowledge, and yet human folly was as pervasive as ever. With the memex and its trails, Bush was offering the tantalizing possibility that a machine could allow individuals to cut through the haze of complexity at a touch of a few buttons, building their own trail of associations as they went. Bush's short essay on the memex suggested a technological fix, and Engelbart's life work has been largely framed in terms of implementing that fix, in terms of using technology to enable knowledge workers to tame uncertainty and complexity.

Engelbart's work was more radical than work in the computing subfield of what was then called information retrieval and that was also an heir to the encyclopedic vision. (Engelbart himself took pains to distinguish his work from information retrieval.) Noting the many differences, Bardini goes so far as to argue that Engelbart was a descendant of Whorfian theories of language, which, if true, would suggest that Engelbart belongs in a non-Cartesian epistemological universe.[53] Bardini does show that Engelbart was indeed aware of the writings of Sapir and Whorf and that, just as Whorf thought that the forms of language could shape consciousness, Engelbart believed that the forms and means of symbolic communication could hinder or help understanding; Engelbart was interested, not just in getting information more quickly or effectively, but in how new structures of communication powered by computers might enable new and better conceptualizations.

Yet Engelbart's project remained one of augmenting *intellect*. It remained, in its stated goals, in its funding, and in its driving assumptions a project of mind.

Whorf, by arguing that different languages engender different conceptualizations of life, rendered conceptualization itself contingent. At least to a degree in Whorf, there are no longer, say, Kantian universal categories of mind but rather contingent forms of understanding rooted in history and everyday life; thinking is of a piece with using language, a language that is inextricably embedded in time and place. As a theorist, Engelbart was not fully Whorfian. His starting assumption was that the problems of the world were ones that called for simply more and better intelligence, not, say, different social relations or an embrace of the contingencies of history and tradition.

Like the work of many nineteenth-century utopians, there's something almost poignant about the sincere idealism in Engelbart's writings, using his turgid, mechanistic engineer's style to describe systems for the betterment of society. Coupled to his foresightedness and modest demeanor, it seems almost churlish to cast a critical eye on his project. Yet it is worth pointing out that the very idealism, the stubborn attachment to his vision, that makes him so appealing tends to lead the discussion away from certain hard questions. On the one hand, the Engelbartian vision contains a deep impatience with conventional scholarly techniques and institutions, with the existing worlds of knowledge. Are the traditional ways of organizing information really all that bad? The sense of utopian hope in Engelbart rests on the argument that computers can provide, not just, say, convenient access to library catalogs and indexes, but a radical improvement on them, one that will augment the intellect. To believe the early Engelbart one has to agree, not only that computers might accomplish the task of easy, intuitive access to and manipulation of information, but that they can do so in a way dramatically better than the elaborate institutions and technologies that have been developed for that task over the last five centuries.

On the other hand, what about Engelbart's version of the encyclopedic dream itself? Is poor access to and control over information really such a central cause of our problems today? Is lack of knowledge really the problem, the key impediment to progress? A key Enlightenment conceit was that once enough information about the world was made available and people were given the means to make sense of it the scales would fall from our eyes, and human society as a whole would be greatly improved. While there's no denying that the scientific revolution has changed human existence dramatically, we also have a century or so of experience that casts into doubt the belief that more knowledge leads automatically to more humane and more reasonable behavior. As exiled German critics Max Horkheimer and Theodor Adorno wrote in 1944, in the full glare of the many scientifically powered horrors of World War II, "the Enlightenment has always aimed at liberating men from fear and establishing their sovereignty. Yet the fully enlightened earth radiates disaster triumphant."[54] Is it better access to information via computers or by other means really going to get at the source of our woes? Are the problems of the world really ones of inadequate intellect, or are they more about, say, social structure, or values, or access to resources?

These are large and difficult questions. The key point here is merely that Engelbart and his followers did not ask them. Engelbart's vision rests on an unreflective acceptance of both an Enlightenment conceit about the need for more knowledge and a sincere belief that computers can enact that conceit where traditional means have failed. And, on the surface, it is a rather dry vision, in the same category as the Dewey decimal system or double-entry bookkeeping; it is possible to wax enthusiastic about the profound importance of these practices (and people have done so), but it is not a vision that on its own terms is likely to generate much passion.

Beyond the Intellect:
The 1968 Demo and the Desire for Interactivity

But, amongst a small but hugely influential group, Engelbart's work did indeed generate passion. Engelbart's work was, and in a certain way remains, somehow thrilling in a way that similar work was not. Engelbart's rationalism was perhaps not the most compelling aspect of his project. This deserves explanation.

One of the more incisive criticisms made of rationalism by early nineteenth-century romantic philosophers was that the rationalists assumed too much. By trying to hitch their analyses to grand, universal, and mathematically specifiable frameworks of Newtonian physics, the Enlightenment rationalists assumed a world of mathematical certainties driven by billiard-ball-like causality even in areas where such certainties were not known to exist, such as the world of human affairs. Everything, they thought, is part of a fixed grid of causes, discernable if one could only shine the light of science, of rational intelligence, brightly enough to cut through the murk of history. And if that is the case, then the solution to problems can be found by breaking them down in terms of events' places in a discoverable causal chain. And from that one can conclude that it is both possible and useful to separate means from ends; if causality follows universal principles, breaking down the specifics into predictable—and therefore instrumentally controllable—processes makes sense.

But, the romantics countered, in some areas of life that grid is nowhere to be found. The romantically influenced philosophers, from J. G. Herder through his many followers such as Coleridge and Matthew Arnold, tended to focus precisely on phenomena driven by internal, nonuniversal, immanent logics; Herder's argument with Kant was that poetry, language, or cultural mores, for example, were driven by unique peculiarities that could not be easily shoehorned into a universal rationality, into a mathlike grid. Partly the argument was empirical; the differences between French and English were accidents of history and could not be explained in term of universals, for example. But the critiques also took the form of alternative understandings of causality: French was French and English was English because of immanent processes, patterns internal to themselves, patterns that could not be understood by mapping them onto chains of cause and effect that followed universal rules. Language or poetry could only be understood from within, as self-driven phenomena.

Engelbart's justifications were on the surface doggedly rationalist, but what he actually tried to build was *self-driven*, a computing environment driven by processes internal to itself. This is most explicit in Engelbart's concept of bootstrapping, elegantly analyzed in Thierry Bardini's wonderful book on Engelbart with that term as its title. In *Bootstrapping*, Bardini shows how Engelbart, in contrast to others in the computing field, had a much richer sense of the importance of social relations in technology development. Engelbart viewed his project as not just one of computer design but as essentially a project of social experimentation involving computers, where those developing his system would be learning to interact with one another over it at the same time that they worked on the system; this in theory would occasion a feedback loop that would steadily improve the system, making it ever more practical. (Van Dam has credited Engelbart with the now common idea of shared software tools, where instead of writing complete programs designed for specific tasks, numerous connectable bits of code—tools—are created so that eventually a programming environment is cultivated that allows large numbers of programmers to build on and take advantage of each others' work; the notion of software tools essentially brings social relations explicitly inside the process of engineering.)[55] Bardini also shows how for Engelbart users were imagined, not just as disembodied minds, but as embodied people who experienced the world in the first instance through the senses, visually but even more importantly through physical touch; this helps explain Engelbart's invention and fascination with things like the computer mouse and the chord keyboard. And Bardini elegantly traces the influx of 1960s countercultural influences in Engelbart's project, particularly as it developed (and eventually degenerated) in the 1970s.

So while Engelbart erected his efforts on a base of classic Enlightenment terms, encounters with his project offered the *experience* of immanent processes in the context of computing. This is perhaps best illustrated by looking at Engelbart at his most influential, the 1968 "mother of all demos."[56] By all accounts, a breakthrough moment in the effort to move beyond batch processing came in December 1968, when Engelbart for the first time publicly demonstrated his system involving networked computers using graphical, windowed interfaces and computer mice to collaborate remotely across a computer network. Many then-young computer scientists who would later go on to revolutionize the way people used computers were present, such as Alan Kay and Andries Van Dam. They both cite Engelbart's presentation as a galvanizing moment. Stewart Brand, creator of the *Whole Earth Catalog* and the man who in the 1970s and 1980s would play a key role in shepherding the computer counterculture into organized existence, manned a camera.[57]

The demo has been widely discussed and celebrated.[58] Yet the exact nature of the thrill of the demo deserves consideration. To a room full of inhabitants of the frustrating world of batch processing, the demo was a window onto longed-for new possibilities. The demo began by displaying, in a full screen, a fictional grocery shopping list. At a time when even the relatively rare interactive terminals generally worked one line at a time, the simple fact of seeing a full-screen display of text, all accessible at once, must have been exciting enough. But Engelbart then proceeded to use the mouse, keyboard, and chord handset to fluently manipulate the shopping list, breaking the list into categories, displaying it as indented sections in outline form, and then—with what must have been jaw-dropping ease given the state of computing at the time—showed the list items linked to points on a graphically displayed simplified map. To an audience already bitten by the computer bug, by the desire to interact with computers, the effect of Engelbart gracefully maneuvering through this online world from his console must have been riveting. It was a world of text and ideas in graceful athletic motion, suggestive of any number of possibilities.

Yet it was merely suggestive. Engelbart and his colleagues did not actually demonstrate any progress on the base claims of his project; beyond work on the system itself (which was in any case still very much a work in progress), there was no evidence of real-world problems being solved, of real complexities being managed or overcome. Even if developed to the point of practicality, an automated shopping list as described by Engelbart would be at best a convenience. To those with specific institutional tasks in mind, like Pentagon officials interested in controlling far-flung military efforts, or Xerox chairman McColough with his interest in Taylorizing the office, the value of Engelbart's work was invisible. Where, exactly, was the intellect that was being augmented? What real-world complexities were being discerned and overcome? This was thrilling only if you were susceptible to the idea that working with computers might be a compelling activity *for its own sake*, without a predetermined goal in mind.

Engelbart began his presentation with the following, now legendary, words: "If in your office, you as an intellectual worker were supplied with a computer display backed up by a computer that was alive for you all day and was instantly responsive, how much value could you derive from that?" The appealing vision of "a computer that was alive for you all day and was instantly responsive" must have seemed extremely tantalizing to this group, most of whom were struggling with batch-processing systems and thus had had only small tastes of direct, "playful" interaction with a computer. Viewers of the demo were treated, not to the abstract principle of solving complex social problems, but to a lively enactment of what it could feel like to interact with an "alive" and "instantly responsive" computer system in an unscripted way, in a way where the interaction was driven by its own internal logics and processes.

The thrill of the demo, in sum, was not about achieving a prespecified goal. The thrill was all about the means. The lack of any specific ends was part of the appeal. It was deep play, an event that exceeded rational justification, that in its celebratory intensity became a kind of community metacommentary, a story not for outsiders, but one that people "tell themselves about themselves."[59] As with Licklider, the desire for interaction with the computer, for that feeling of "self-motivating exhilaration," was seeking and finding a justification, a shared meaning. The demo succeeded, particularly insofar as it offered its attendees not just ideas, but, in the audience's enthusiastic reception, confirmation and encouragement of the connection of the desire for interactivity with a sense of direction for computing. After the Engelbart demo, individuals with an interest in computing had a new, grand way to make sense of and justify their own desire to interact with the computer. And in the standing ovation that concluded the presentation, they could know, they could *feel*, that there was a community of others who shared their convictions. When they returned back to their university and corporate laboratories to labor away at the limited machines of the day, they had a new sense of who they were as they worked, a new meaning for the act of working with and programming a computer.

Conclusion

Among the still small community of computer professionals in the 1960s, computers attached to screens and keyboards were working as thought objects; the compulsive fiddling they occasioned elicited, not only a desire for more, but a search for intellectual frameworks to justify that interaction and extend it. Interactive computing was occasioning practices that pushed against the boundaries of the instrumental reasoning that put the computers there in the first place. From within a world in which computers were assumed to be tools with which leaders have things done, such as directing remote battles, Licklider and his ilk were developing a fascination with interactive computing's "self-motivating" qualities, and that fascination pushed them toward other logics, toward the idea of using computers to formulate questions rather than answer them, toward understanding computers as tools for exploration through symbolic manipulation rather than for conquering known territories or organizing human affairs into a predictable grid. Licklider and Engelbart's efforts are symptomatic of how a felt experience—computer holding power—became the occasion for the search for new frameworks of meaning for making sense of that experience, frameworks that would eventually come to shape the development of the machines themselves. Noninstrumental visions of computing, where activity was "self-motivated" rather than strictly goal-directed, were emerging from within an institutional framework heavily dominated by forms of instrumental reasoning that separated means and ends.

The significant point here is that personal experiences with interactive computing were shaping the development of new ideas as much as new ideas were driving new uses of computers. Neither Licklider's nor Engelbart's writings fully add up as empirical or philosophical statements; Licklider oddly mixes cold war instrumentalism with gestures towards other logics, and Engelbart's demo offered its audience a way of making sense of computing that exceeded Engelbart's theorization of it. Engelbart painted a picture of a pure mind-world, a neat, hierarchically organized domain inside the computer screen, yet what his audience walked away with was a strong sense of possibility about what interactive, non-goal-directed computing could be like, whether or not the end point of augmented intellect was ever reached. And the demo may have worked for his audience because most of them already had had at least small tastes of interacting with a computer. (It may have been the physical presence of others in the audience with a shared experience—the compulsive desire for computer interaction, and the knowledge that one was in the physical presence of others who shared that desire—that made the demonstration so compelling; after the demo, the compulsion to interact with the machine no longer needed to be seen as a random oddity or weakness; being present at the demo gave one a new set of publicly available meanings to attach that experience to.)

In sum, what one sees in the 1960s is the complex interaction of big ideas—cold war and managerial forms of instrumental reasoning—with actual face-to-face experiences with technologies, technologies that the big ideas paid for and informed, in ways that sometimes reinforced each other but also sometimes created tensions between experience and formal plans, tensions that hinted at alternate directions. In the next decade, as other habits of thought fueled both by social ferment and longstanding traditions swirled through the culture, those new habits of thought would also in their own way interact with the developing computer world and leave their imprint on the ways in which computers were imagined and built.

2 Romanticism and the Machine
The Formation of the Computer Counterculture

> Whoso would be a man must be a nonconformist.... Trust thyself ... that
> science-baffling star, without parallax, without calculable elements, which shoots
> a ray of beauty even into trivial and impure actions ... that source, at once the
> essence of genius, of virtue, and of life, which we call Spontaneity or Instinct.[1]
> —Ralph Waldo Emerson

Introduction: Reenchantment

In 1972, four years after the Engelbart demo, Stewart Brand penned an article for *Rolling Stone* entitled "Spacewar: Fanatic Life and Symbolic Death among the Computer Bums."[2] Visiting Xerox PARC just as its engineers were further developing some of Engelbart's concepts, Brand had decided, not just that computer programmers now sometimes sported long hair and sandals rather than the crisp white shirts, crew cuts, and black ties associated with IBM engineers, but that some of them were exploring an approach to computing that was something quite different. "The general bent of research at Xerox [PARC is] soft," Brand wrote, "away from hugeness and centrality, toward the small and the personal, toward putting maximum computer power in the hands of every individual who wants it." Brand's article now looks quite prescient; it predicted, for example, the decline of record stores in the face of computer-network-delivered music and quoted many of the now-legendary heroes of the early stages of computer revolution. But, most significantly, it pointed towards a different set of cultural associations for computers and computer programmers.

Brand announced this radical revision of the meaning of computers in his title, with both the name of a computer game and a nod towards Jack Kerouac's novel and alternative lifestyle manifesto, *Dharma Bums*. By putting a game at the center of the article, Brand presented computers, not just as liberating, but as fun, and perhaps liberating *because* they were fun. This was not Engelbart's Enlightenment vision of personal computer use for the serious purpose of solving complex social problems. Brand did not discuss business or educational applications, or project visions of libraries of the future, or potential new efficiencies in scientific research. Rather, Brand was linking the experience of "self-motivating exhilaration" to a creative kind of pleasure. If Engelbart was for the most part trying to use computers to enact an electronic version of Diderot's encyclopedia, Brand, like much of the counterculture, was building an association with something that might best be called Byronic—a traditionally *romantic* sense of pleasure that mixed rebellion with a sense of individual creativity and expression.

Romanticism and Modernity

Faced with the dull weight of a highly specialized, technologically and bureaucratically organized world, at various moments, many of us go off in search of ways to bring elation to our lives, to bring the magic back, to recover what Max Weber described as enchantment; and on this quest we expend energy, careers, lives. Sometimes this impulse simply peppers the social fabric of industrialized societies and comes out in random instances of individuals suddenly turning to, say, mountain climbing or abrupt changes of careers or spouses. But at times the impulse becomes organized and can lead to paroxysms of social change, such as the diverse religious movements that currently convulse our world.

If religion is one form the search for reenchantment can take, another is romantic individualism. As a concept, Ralph Waldo Emerson's "Self-Reliance" provides a concise summation. "Whoso would be a man must be a nonconformist," Emerson argued. Resist conformity and consistency and instead, "Trust thyself," where the self is understood as "that science-baffling star, without parallax, without calculable elements, which shoots a ray of beauty even into trivial and impure actions . . . that source, at once the essence of genius, of virtue, and of life, which we call Spontaneity or Instinct."[3] This is not Descartes's rational self and certainly not the classical economist's calculating shopkeeper self; it is a self defined exactly against calculation and predictability. Emerson proclaimed the centrality of a dynamic, inner experience that calls on us to live creatively beyond the bounds of predictable rationality, to express ourselves according to our own unique personal perception of truth.[4]

But romanticism is more than just a concept. The literary scholars who use the term most systematically tend to classify romanticism in terms of specific great authors and associated texts; in this sense, romanticism is understood as a collection of great works or as a period in European history usually placed in the early nineteenth century.[5] Yet for the last two centuries or so, people who have never read Emerson, Wordsworth, Byron, or other romantic era hallmarks, have repeatedly produced and consumed tales of revelation based on inner experience, celebrations of art as what Wordsworth called "the spontaneous overflow of powerful feelings," and other characteristic features of romanticism: stories of heroic outcast wanderers on desperate quests, a fondness for iconoclastic ideas presented in authentic-sounding plain language, enthusiasm for the apparently impulsive overthrow of the dead weight of history and tradition. So an alternate understanding of romanticism has been developing with a more sociological slant. Friedrich Kittler suggests that romanticism is best understood as a set of discursive practices available to and distributed widely through the culture at large. Colin Campbell's neo-Weberian theory of the "romantic ethic's" role in modern consumerism is also useful in this regard, though he approaches the question from a different angle than Kittler.[6] The point is that the great romantic authors may have been responding to the romanticism in the culture at large, rather than creating it. The object of analysis is not in texts, but in society; texts are simply one way of getting access to it. Though it probably first emerged in the late eighteenth century (at least according to Campbell's and Kittler's accounts), in our day romanticism has become a kind of cultural toolkit, a grab bag of cultural habits, available for use in a variety of contexts, from candle-lit dinners to therapy sessions to the 1960s antiwar counterculture.[7]

It was in this sense that, in late-twentieth century America, romantic individualism became attached to, and came to have an impact on, computer networks and their place in our world.

Understanding the 1960s Counterculture and Its Legacies

The decisive history of the relation between the 1960s counterculture and the development of a computer counterculture is Fred Turner's *From Counterculture to Cyberculture*.[8] Turner points out that there was never a single, unified counterculture. It was something composed of several different strains of thought and practice. While the New Left strain carried with it a call to engage the political apparatus in a broad way, Turner focuses on another distinct strain that he calls the "New Communalists" who instead of engaging the political system sought to escape it by transforming consciousness and creating autonomous communities devoid of hierarchy or rules. It is this New Communalist strain, Turner argues, centered around Stewart Brand and the *Whole Earth Catalog*, that embraced a particular vision of technology and that would eventually provide the intellectual underpinnings of the cyberutopian movements of the 1990s.

Turner has elegantly demonstrated several key points. First, the 1990s cyberculture had very strong continuities with the 1960s New Communalist movement, both in terms of individual participants like Stewart Brand and in terms of intellectual frameworks and practices. Second, in an important contribution to the sociology of knowledge, Turner shows how most of this rested on a practice of creating "network forums" in media, think tanks, and conferences that build rhetorical and practical bridges allowing collaboration between diverse communities of thought, such as the military-industrial-complex and artistic communities. Third, Turner shows how the New Communalists, for all their egalitarianism, also embodied a contradictory tendency towards a kind of elitism of charismatic insiders, a product of their setting of a boundary between those in the know and those fallen souls of the old ways. "Like the communards of the 1960s," he writes, "the techno-utopians of the 1990s denied their dependence on any but themselves."[9]

Once it is established that a pattern repeats itself across several decades, the question raised is, what sustains it? Why did the New Communalist rhetoric and practice endure? What is its appeal across time and contexts? Here I focus on the specifically romantic individualist character of the counterculture as it migrated into the world of computing, how it stands out against a backdrop of various kinds of social disaffection, and how it provided ways to make sense of and legitimate the embodied experience of compulsive computer use.

Social Disaffection and Shifting Visions of Computing in the 1960s

That moment of cultural and political turmoil generally referred to as the sixties (which actually occurred roughly between 1964 and 1972) had the disaster of the Vietnam War at its center. Paul Edwards offers Operation Igloo White as the archetypal example of the horrifying collision of 1960s systems-science-influenced computing with the complex realities and passions of the time. Operation Igloo White was a giant computer-based command-and-control system based in Thailand during the Vietnam War that gathered data from electronic sensors hidden along the Ho Chi Minh trail and then used that data to direct near-instantaneous bombing strikes in the jungle. Soldiers on the ground were supplanted by men in far away closed buildings staring at computer terminals. It was a case study in hubristic folly; while causing large numbers of casualties on both sides, the effort did not stop the effective military use of the trail, and many now argue its main function was to help blind overconfident U.S. military leaders to the reality of the situation on the ground.[10]

But the 1960s would not have been what it was without countless small struggles and shifts throughout the social fabric, shifts that occurred in work places and homes lasting well into the 1970s. Those struggles manifested themselves in the worlds of computing on several levels. For example, the centralized information processing championed by both IBM and Xerox's upper management began to run aground on the shifting sands of social expectations about work and gender. By the mid-1970s, IBM's earliest push for centralized "word processing centers" was accused of having caused some "disasters" and to have produced conditions that some executives recognized as "dehumanizing." The feminist movement comes up surprisingly frequently in the business press of the time; an executive told *Business Week* that part of the problem with sequestering secretaries in word processing centers was that "no one used to worry about the career path of secretaries. . . . Women's lib is affecting things now." Another executive observed that IBM's plan "flies in the face of the team concept and women's lib."[11] Upper management's dream of a docile, factory-like efficiency in the office was backfiring, generating friction and resentment instead of cost-savings and profits. As a result, up and down the corporate hierarchy, as digital computers continued to spread, more and more individuals were having experiences that suggested there was something wrong with the dominant ways of thinking.

In other arenas, computer-related visions that in the mid-1960s seemed to hold much promise were similarly colliding with recalcitrant realities. Systems science, for example, began to loose its high-tech sheen by 1970.[12] After several years of city governments pouring money into consultants and computer systems, urban crime and strife were increasing, not decreasing, and the idea that computer systems might be the way to tame the unruly complexity of urban life was looking increasingly naive.[13] At roughly the same time in academia, critics (such as Herbert Schiller in his 1969 *Mass Communication and American Empire*) began to point sharply to a gap between the goals of democracy and modern communication systems.[14] James W. Carey theorized the issues in the early 1970s. In "The Mythos of the Electronic Revolution" (1972), he attacked systems-science-influenced approaches to communication, making the case that, for all their scientific pretensions, the arguments were actually pseudoreligious. His seminal 1975 essay "A Cultural Approach to Communication" made explicit the distinction between communication as control on the one hand and communication as something more horizontal on the other. The essay's core distinction—between a "transmission" view of communication with a focus on control-at-a-distance (and which Carey blamed for the "chaos of modern culture") and a "ritual" view with its focus on the generation and maintenance of shared meanings—has been a staple of academic discussions of communication ever since.[15]

The common sense of the culture was not just growing more skeptical of the more grand expectations of computers; the basic terms through which they were understood were shifting. In 1961, Licklider could easily lump together the use of computing for winning battles with the use of computers for achieving enlightenment. By 1969, that juxtaposition was no longer easy or friction-free. The use of computers for control and the use for communication and expression no longer seemed so obviously compatible and were coming to be experienced as in tension with one another.

We are now famously aware that when Xerox chairman McColough was enthusiastically imagining the office of the future in terms of efficient centralization, a handful of his employees were busy at Xerox's Palo Alto Research Center in California creating some of the first effectively functioning computers with windowing graphical interfaces, mice, and network capabilities. McColough and the rest of Xerox's upper management failed to understand the value of these innovations and thus largely ignored them and allowed them to be copied first by Apple's Lisa and Macintosh systems and eventually by Microsoft's Windows. It is now known as one of the bigger business blunders of all time and has been much discussed elsewhere.[16] But what's significant here is that the difference between Xerox upper management's view of things and the PARC engineers' view was to a large degree about the relation between communication and control. The problem was not that Xerox's upper management was unintelligent but that they understood computing and data processing through the dominant 1960s lens; like the early military funders of ARPA's computing efforts, they imagined computer communication as a means of control over people and events at a distance, as tools that would bring mathematical certainty into the pink-collar world of the front office, as a form of efficient control across space. The PARC engineers, by contrast, were carrying on Engelbart's and Licklider's move towards an alternative view, viewing computers as less calculators than symbolic manipulators, and computer communication as something that should give more control to users rather than giving management more control over them. The dominant point of view was not that of a manager or military officer overseeing a vast enterprise; it was that of literate individuals achieving goals with an interactive computer under their own control.

But the failure of Xerox's upper management to foresee the direction of computing in the 1970s occurred in a broader social context. The cultural and political crisis precipitated by the Vietnam War had its effects on the community of computer engineers and visionaries, just as it did on so many other aspects of the society. By the late 1960s, the unquestioned enthusiasm for cold war militarism that had previously provided much of the cultural glue binding together the military-industrial complex began to weaken; like many intellectuals with university associations, substantial numbers of computer engineers and scientists began to be influenced by the political currents associated with the counterculture. Most histories of computing make some reference to the symptoms of this: the programmers working on ARPANET who wore sneakers and antiwar pins to briefings at the Pentagon in 1969; the appearance of a very early email message in 1972 on ARPANET—which was then understood as primarily a military communication system—calling for the impeachment of President Richard Nixon.[17]

Did it make a difference that the pioneering PARC engineers and scientists had a famously relaxed, countercultural style, that they had meetings in rooms full of bean bag chairs instead of conference tables, that they dressed informally? By themselves, probably not; but those small stylistic differences were symptomatic of broad cultural shifts in the society at large, shifts that taken together helped change the center of gravity in the dominant visions of what computers were for and how they might be built and used. By the early 1970s, it had become easier to think of computing in ways that were not congruent with the demands of the cold war era.

Humanist Romanticism

There was more than one alternative to cold war militarism. One response to the tensions of the time is evidenced in MIT computer scientist Joseph Weizenbaum's 1976 book, *Computer Power and Human Reason*, a sweeping critique of the use of computers in American society.[18] Like his colleague Norbert Weiner before him, Weizenbaum, midway in a successful scientific career, had grown concerned over the indifference of the scientific community to the destructive uses of their discoveries and inventions. If Weiner's archetypal scientific sin was the atom bomb, however, Weizenbaum wrote in the shadow of the Vietnam War. And, in the case of Vietnam, the problem was less about horrifically effective weapons and more about a kind of structured blindness or indifference that enabled horrors to be committed using conventional weapons. Weizenbaum cites, for example, the use of computer systems, "operated by officers who had not the slightest idea of what went on inside their machines," to select "free-fire zones" within which "pilots had the 'right' to kill every living thing." And he cites the notorious case of Pentagon computers that listed bombing strikes inside Cambodia as occurring in Vietnam, thus using the mystique of the computer to mislead members of Congress about this arguably illegal action.[19]

As a computer scientist, Weizenbaum was not critical of computers themselves, but of what he saw as a general weltanschauung that had become associated with computers. Yet Weizenbaum's criticism was broad and targeted at concepts towards which some of his MIT colleagues had devoted their careers. He saw the notion that human beings could be understood along the model of computers—the core conceit of the field of artificial intelligence—as of a piece with the narrowness of mind, the instrumental reasoning that separated means from ends, and the inhuman grandiosity that seemed to be associated with the computing weltanschauung.

Weizenbaum's critics sometimes dismiss him as a Luddite.[20] He is not. Yet readers might get this impression, not just from his expression of doubt about certain aspects of computing, but because of his loosely romantic understanding of human creativity, which through the 1970s was for the most part associated with a vision of nature set against technology. Weizenbaum began his book with what he calls the "obvious idea" that "science is creative, that the creative act in science is equivalent to the creative act in art, that creation springs only from autonomous individuals."[21] The assumption that the archetype of creation is artistic creation and that such creation "springs only from autonomous individuals" is not a Luddite one, but it is an orthodox element of romantic individualism. And, more often than not, romanticism is thought of as antitechnological because of its privileging of a vision of nature contrasted with a demeaning, rationalizing industrialization. From Wordsworth finding truth in the natural simplicities of the English countryside to Thoreau in his cabin at Walden to 1960s hippies building communes on farms in Vermont, romantic movements often define themselves against what they perceive as the blinkered technical rationality of the industrial world, which they often contrast with agricultural forms of life. By the first half of the twentieth century, this strain of thought had been fully developed in the work of humanist critics like Jacques Ellul and Lewis Mumford (the latter cited by Weizenbaum as an important influence) who attributed many of the ills of the modern world to the forms of consciousness associated with technology.

Computer Power and Human Reason indeed belongs in the Ellul/Mumford tradition, principally because of its romantic understanding of the human as in essence creative and because of its Hegelian analytical method of identifying a central essence or "spirit" in culture that is expressed throughout the society. Yet his critics' accusation of Luddite is both on its face inaccurate—Weizenbaum made clear he was not criticizing computers per se, just particular uses of them and specific actions taken in their name—and tellingly defensive. For the most compelling and unique parts of the book are in Weizenbaum's specific discussions of the workings of computers and the activities of those who make and use them.

Weizenbaum's chapter called "Science and the Compulsive Programmer" was one of the first published works to directly address "hackers" and the felt experience of a certain kind of computer programming. Basically, Weizenbaum labels Licklider's "self-motivating exhilaration" as an addiction akin to gambling. As if in retort to Stewart Brand's "Spacewars" essay, Weizenbaum observed,

> Whenever computer centers have become established, . . . bright young men of disheveled appearance, often with sunken glowing eyes, can be seen sitting at computer consoles, their arms tensed and waiting to fire their fingers, already poised to strike, at the buttons and keys on which their attentions seems to be riveted as a gambler's on the rolling dice. When not so transfixed, they often sit at tables strewn with computer printouts over which they pore like possessed students of a cabalistic text. They work until they nearly drop, twenty, thirty hours at a time.

Their food, if they arrange it, is brought to them: coffee, Cokes, sandwiches. If possible, they sleep on cots near the computer. But only for a few hours—then back to the console or the printouts. Their rumpled clothes, their unwashed and unshaven faces, and their uncombed hair all testify that they are oblivious to their bodies and to the world in which they move. They exist, at least when so engaged, only through and for the computers. These are computer bums, compulsive programmers.[22]

The difference between the compulsive programmer and "a merely dedicated, hard-working, professional programmer," Weizenbaum writes, is that the professional "addresses himself to the problem to be solved, whereas the compulsive programmer sees the problem mainly as an opportunity to interact with the computer. . . . The professional regards programming as a means towards an end, not as an end in itself."[23]

Whereas Licklider tried to justify his pleasure in playing with computers by associating it with goals of efficiently solving problems, Weizenbaum—who later hinted that his depiction of programmers was based in his own personal experience with computers[24]—saw it as symptomatic of much that was wrong with society. Programming without concern for the final ends exemplified the indifference to long-term human consequences that lead to destructive folly like massive bombing campaigns in Vietnam. Prefiguring Edwards's argument about the "Closed World," Weizenbaum argues that the appeal of computers is precisely that they offer a fantasy of abstract, solipsized "worlds" shut off from the messy complexity of reality, worlds that can be made to operate according to any rules that the programmer wishes; the programmer gains a kind of godlike control over this world, but only insofar as he or she disconnects from the larger, human one. For Weizenbaum, the compulsive desire to interact with computers enabled the mindless tinkering—the technological innovation without regard for human consequence—that was the object of his concern. Hacking was the opposite of art.

"Soft" Computing:
The Emergence of the Computer Counterculture

For Stewart Brand, however, computing potentially *was* art. While Weizenbaum was formulating his critique of conventional computer science, a computer counterculture was developing which offered a parallel critique of the computer establishment, but one that took an opposite tack precisely on the issue of "compulsive programmers." Stewart Brand's aforementioned essay in *Rolling Stone* was both indicative and probably influential. Brand took the same raw material as Weizenbaum—computing conducted compulsively for its own sake—and articulated it completely differently, as something that might free us instead of perpetuate our enslavement. Brand's "computer bums" were not portrayed as Weizenbaum's lonely addicts but as visionary computer beatniks.

This move was not without its intellectual deep background. Not all of the attendees at the 1940s Macy conferences followed Norbert Wiener and Licklider into the world of computer science; anthropologists Gregory Bateson and Margaret Mead were also in attendance.[25] Bateson in particular would continue using the term *cybernetics* into the 1970s in ways that would come to seem quite at odds with the Strangelovian "Closed World" later described by Edwards. Bateson, who to my knowledge never cared much about computers, went on to develop both a set of holistic ideas about systems theory, ecology, and the human mind and a particularly effective, aphoristic pop writing style for presenting those ideas. Bateson's trade books from the countercultural period, the most famous of which was *Steps to an Ecology of Mind*, were written in a highly accessible, engaging way that eschewed academic jargon and reference; the style was that of a kind of hip, charming version of the voice of the British gentleman amateur. Highly abstract ideas about systems theory, for example, are put in the mouth of a six-year-old girl chatting with her father.[26] Hence, college students and literate hippies across the land, and even some precocious high school students, could curl up in a bean bag chair with one of Bateson's books and make some sense of it without the guidance of professors; Bateson was an anti-Derrida.

This was the intellectual style that became most powerfully associated with the counterculture, particularly its New Communalist wing: accessible, smart but plain-spoken, dismissive of tradition. Someone like Weizenbaum would make frequent references to classic literature like Dostoyevsky and project a sense of warning about a fallen, deluded world. Bateson, by contrast, would boil things down to easily repeatable aphorisms and provide a sense of revelatory simplicity.

Stewart Brand's *Whole Earth Catalog*, the defining text for the 1960s counterculture's approach to technology, adopted much of Bateson's style and eventually elevated him to the status of guru. In the *Catalog*, Brand added to Bateson's accessible but thoughtful style a nonlinear, playful form of presentation that mixed descriptions of nonflush toilets with political tracts, a novel, and iconoclastic journalism; it was in the *Catalog* that most of the United States finally learned how astronauts went to the bathroom. On the one hand, the style expressed the "everything is related" holism of Batesonian systems theory. But the *Catalog* was also made for browsing. Certainly, the accessible, cluttered style of the *Catalog* shared something with the general style of the consumer culture; reading the *Whole Earth Catalog* in the early 1970s was probably fun in much the same way that browsing the Sears catalog was in the 1890s. But when it first appeared, the *Whole Earth Catalog* stood apart from the rest of the consumer culture in important ways; printed in black and white with grainy images, it was information rich, deliberately lacked glitz, and was not about consuming products for leisure time activities but—in its own mind at least—about understanding and building things for everyday life. To a whole generation of readers, and still to some extent today, this kind of writing is a breath of fresh air; its frankness and thoughtfulness was an antidote to the breezy, sugarcoated, condescending, anti-intellectual tone of much of the pop media, whereas its accessibility contrasted with the jargon-ridden, mystified styles that permeate our academic, government, and corporate bureaucracies.

It was probably the *Whole Earth Catalog* that disseminated many of the tropes now emblematic of New Communalist countercultural writing. A studied use of plain, conversational language was common, signaled by the artful occasional use of obscenities, more for humor than to shock or express anger. The *Catalog's* statement of purpose begins, "We are as gods and might as well get used to it." The use of conversational plain language—"might as well get used to it"—often functioned to humanize what would otherwise come across as almost biblical grandiosity: "We are as gods." In this way, undeniably grand statements about abstractions, about mind, society, the "whole earth" could be presented in a disarmingly appealing way.

Ted Nelson and Computer Lib/Dream Machines

If the *Whole Earth Catalog* brought a new take on technology to the counterculture generation, Ted Nelson's *Computer Lib* did the same for computers, with lasting impact. Among computer enthusiasts, Nelson was the central figure in building an association between interactive computing and countercultural style. He coined the term *hypertext* in the early 1960s and in 1967 coauthored a proposal for a computer editing system with Brown professor Andries Van Dam, a friend from Nelson's undergraduate days (and one of the attendees who would be so deeply impressed by Engelbart's 1968 demo).

Nelson has never been particularly successful either technically or in business; to my knowledge, no functioning software or business with which he has been directly associated has ever endured. Arguably, his key role in the evolution of computing—and this is not for a second to downplay his influence—is in his role as a writer. Nelson is a magnificent and distinctive prose stylist, and in a sense his career is proof of the power of the literary to change the world. Although Nelson was clearly aware of and influenced by the likes of Engelbart and Licklider, by the early 1970s his tone and style became enthusiastically countercultural. Licklider and Engelbart, though they were somewhat interested in using computers playfully, generally used a language inflected by the turgid technocratic talk characteristic of military-industrial bureaucracies. In sharp contrast, Nelson created an ebullient and stunningly effective rewriting of the technical into a flip, iconoclastic, amusing, and sometimes poignant style.

As a student and drifting young aspiring professional in the mid-1960s, Nelson published a few pieces and gave some public talks that advocated an Engelbartian emphasis on using computers to create and manipulate linked texts. These early pieces are largely of a piece with the Licklider/Engelbart strain of thought. A 1965 article that contains the first published use of the term *hypertext*, for example, looks at first glance very much like a fairly typical engineering piece, proposing a system for "personal filing and manuscript assembly."[27] In keeping with classic engineering form, Nelson coins a technical term for his proposed system for handling electronic documents, which is easily turned into an acronym: the Evolutionary List File structure or ELF. Like Licklider and Engelbart, Nelson distinguishes his approach from the mainstream of the day by emphasizing the unpredictable quality of mental processes. "Rarely," Nelson writes, "does the original outline [of a project] predict well what headings and sequence will create the effects desired. . . . If a writer is really to be helped by an automated system, it ought to do more than retype and transpose; it should stand by him during the early periods of muddled confusion, when his ideas are scraps, fragments, phrases, and contradictory overall designs."[28]

While Nelson mentions potential uses that echo Licklider and Engelbart, like scientific or legal research, he pushes things a bit farther, suggesting that this system could be used by historians or students of Shakespeare.[29] He hints at a fondness for the playful non-Latinate language by proposing a system he calls "zippered" lists. More than Licklider or Engelbart, Nelson is focused on the vision of an individual who is, not just managing data, but *writing*. "To design and evaluate systems for writing," Nelson asserts, "we need to know what the process of writing *is*."[30] And writing, according to Nelson, is not simply typing words into a system, nor is it a predictable process of, say, following a pregiven outline. Writing is a process of gradual discovery that involves "balance of emphasis, sequence of interrelating points, texture of insight, rhythm, etc." In a footnote, he continues, "I understand that this account is reasonably correct for such writers as Tolstoy, Winston Churchill and Katherine Anne Porter. Those who can stick to a prior outline faithfully, like James Fennimore Cooper, tend to be either hacks or prodigies, and don't need this system."[31] Neither Licklider nor Engelbart could ever have produced a sentence like that. Both the literary references and the artful rhetorical dodge regarding Cooper point to something that distinguishes Nelson from Licklider and Engelbart; he is fascinated with, and skilled at, the craft of writing.

As the 1960s progressed, Nelson came into contact with various avant-garde artists who had been challenging the typical assumption that art and engineering were opposites. Composer John Cage, for example, besides creating controversial compositions (perhaps most famously 4'33"), was fascinated with technology. Kathleen Woodward has contrasted Weizenbaum's inspiration, Lewis Mumford, with John Cage: "If for Mumford, the values associated with technics are impersonality, regularity, efficiency, and uniformity, for Cage the values are heterogeneity, randomness, and plenitude. These are also the values he associates with art."[32] This line of thinking about technology and art led to a number of European and New York-based efforts. For example, Billy Klüver, Robert Rauschenberg, and others founded Experiments in Art and Technology in 1966, which held exhibitions in and around New York City in subsequent years.[33] In 1970, Nelson himself wrote an online, hypertext catalog for an exhibition titled "Software," which involved computer- and electronics-related installations, some of which were created by artists, others by engineers, and some who straddled both worlds, like Nicholas Negroponte, then head of MIT's Architecture Machine Group.[34]

Nelson may not have been central in these circles, but he most likely gained some inspiration, of two kinds. First, these groups encouraged him to attack the assumption that art and computer technology were opposites. Second, they perhaps suggested an alternative career path. Instead of becoming a traditional computer expert working for corporations, universities, or the government, someone might develop a name for themselves as a gadfly or iconoclast at the intersection of some of these institutions; one could operate via what Fred Turner calls "network forums" that used the terminology and authority of one world to build links with another. Towards this end, the persona of the artist-rebel and the authority of high technology might be combined instead of being thought of as opposites.

While rethinking the relation between art and technology, however, the New York-based experimental artistic community still remained arch-modernists, expecting art to take risks and be challenging. It was the San Francisco-based counterculture community with Stewart Brand at its center that provided a more popular, inviting approach focused on a warm kind of playfulness and accessible style. And by the late 1960s, whether through rock festivals like Woodstock, new publications like *Rolling Stone*, or simply in the dorms of college campuses across the land, this countercultural ethos was widely available. This became Nelson's primary stylistic influence.

It was two years after Brand's 1972 *Rolling Stone* piece first publicized the notion that computing could be countercultural that Nelson came out with his magnum opus, *Computer Lib/Dream Machines*.[35] While the larger counterculture was at the time very much in decline and paid little attention, *Computer Lib* had a profound impact on the small but energetic circles of people who were, or who might become, involved with computing. For the last two decades, it has been common to encounter computer professionals like Mitch Kapor (the designer of

Lotus 1-2-3 and cofounder of the Electronic Frontier Foundation), who say Nelson's book "changed my life."[36] (In the mid-1980s, Nelson claimed to have encountered at least fifty other people who told him the same thing.)[37] It is impossible to establish exactly how widely read *Computer Lib* was, but it seems likely that most of those in attendance at the West Coast Computer Faire and similar now-legendary venues of the 1970s had at least some familiarity with Nelson and his work, and Nelson himself reports glowingly on a visit to Xerox PARC during the period.[38] Nelson frequently published essays in science and computer journals and served for a time as editor of one of the first pop computer magazines, *Creative Computing*.[39] But it was *Computer Lib* that seems to have had such a formative impact. The book has been in all seriousness described as "the most important book in the history of new media."[40]

Computer Lib was essentially a transposition of the style, format, and countercultural iconoclasm of the *Whole Earth Catalog* into the world of computers.[41] By the time he was working on *Computer Lib*, Nelson had completely abandoned all hints of the technical style of Licklider and Engelbart. Near the beginning of the book, Nelson draws a distinction between the computer professional and the "computer fan," that is,

> someone who appreciates the options, fun, excitement, and fiendish fascination of computers. . . . Somehow the idea is abroad that computer activities are uncreative, as compared, say, with rotating clay against your fingers until it becomes a pot. This is categorically false. Computers involve imagination and creation at the highest level. Computers are an involvement you can really get into, regardless of your trip or your karma. . . . THEREFORE, welcome to the computer world, damnedest and craziest thing that has ever happened. But we, the computer people, are not crazy. It is you others who are crazy to let us have all this fun and power to ourselves. COMPUTERS BELONG TO ALL MANKIND.[42]

The rationalist, technical jargon was gone and replaced by a full-throated counterculturalism: part Tom Wolfe, part Haight-Ashbury, and part political flyer.

Which is not to say that *Computer Lib* lacked substance. Mixed in with descriptions of specific machines and computer languages are concepts and approaches to computer use that were then unusual but have since become commonplace. User-friendly interfaces, small personal-sized computers, mice, graphic interfaces, and noncomputational uses of computers like word processing, email, multimedia, and hypertext are all elaborately explained and advocated. In one of many prescient passages, Nelson attacks a *Business Week* piece about "The Office of the Future," which forecast the emergence of computerized offices staffed by centrally located, specially trained word processing technicians and which predicted that the only companies that will succeed in this field will be IBM and Xerox.[43] Nelson goes on:

> Well, this is hogwash.... The office of the future, in the opinion of the author, will have nothing to do with the silly complexities of automatic typing. It will have screens, and keyboards, and possibly a printer for outgoing letters . . . All your business information will be callable to the screen instantly. An all-embracing data structure will hold every form of information—numerical and textual—in a cats'-cradle of linkages; and you, the user, whatever your job title, may quickly rove your screen through the entire information-space you are entitled to see. You will have to do no programming.[44]

This is an extraordinary bit of prognostication. He even anticipates buzz words; eighteen years before the phrase *web surfing* spread throughout the culture, Nelson wrote, "If computers are the wave of the future, displays are the surfboards."[45] And Nelson articulates grandiose notions about computers' liberatory potential that later became standard fare among netizens, claiming that "knowledge, understanding and freedom can all be advanced by the promotion and deployment of computer display consoles (with the right programs behind them)."[46]

The style of *Computer Lib* shares much with both Bateson and the *Whole Earth Catalog*. The book criticizes and pokes fun at the mystifying jargon in which computers were then typically described. "I believe in calling a spade a spade—not a personalized earth-moving equipment module," Nelson quipped.[47] The language is deliberately playful and non-Latinate; computers are described as "wind-up crossword puzzles." (Like Brand, Nelson frequently uses the colloquial particularly effectively to soften grandiosity, thereby disarming the reader's skepticism: "When I saw my first computer," he recounts, "I said 'Holy smoke, this is the destiny of humankind.'")[48] And a loose sympathy with countercultural politics and iconoclasm is also present; Nelson boasts of having been at Woodstock,[49] associates his critique of the computer profession with the feminist critique of the medical profession in *Our Bodies, Ourselves*,[50] inserts a solemn paean to no-growth economics,[51] and puts a black-power style raised fist on the cover. And the book's hand-drawn graphics, paste-up style, and self-published origin—Nelson brags about eschewing mainstream publishers—all bespeak an antiestablishment sentiment.

In Brand's "Spacewar" piece, the simple fact that people enjoyed playing with computers was an astonishing enough idea. Nelson's *Computer Lib* greatly expands on this, associating "self-motivating exhilaration" (what Nelson calls "fiendish fascination"), not just with play, but with "imagination and creation at the highest level." Nelson was perhaps the first to clearly suggest that computer-enabled virtuality was, not just a system for rational exploration in the Enlightenment sense, but potentially an ecstatically pleasurable activity. Nelson's emphasis on play and personal expression thus allowed for a full break from the stiff Cartesian mechanistic rationality that Engelbart was still rooted in, and his dragon-slaying iconoclastic stance sharply distinguishes him from Licklider's comfortable association with the military-industrial complex. And Nelson pokes fun at the artificial intelligence community as "God builders"; for Nelson, computers are not machines that think on their own but tools that people use to pursue their dreams—"Dream Machines."

Ted Nelson and the Romantic Persona of the Visionary Rebel Hero

How could a slim book about computers change the course of someone's life? No doubt there is something comic about the idea of people sitting at computer consoles imagining themselves as Byronic heroes; one has to approach the notion of romanticized computing with a sense of irony. But one way to make sense of this is to think of romanticism as a social formation, not just an aesthetic or a philosophy. We can think, not just of people like Byron and other romantic figures, but of the readership of Byron, more than a few of whom were in a sense bored bureaucrats, people with relative material security suffering from alienation in their narrow, specialized, and technical professions, dreaming of a different life— looking for reenchantment. One might be able to trace a fairly direct line from some of the earliest masculine heroes of romantic literature—Goethe's young Werther, say—onward to the protagonists of cyberpunk novels, typically mid-level technical employees who've spent a large part of their lives sitting at computer consoles engaged in narrow, technical tasks and then in the course of the story have dramatic adventures.

It's entirely appropriate that Nelson dubbed his proposed ideal hypertext system Xanadu, after the imaginary pleasure palace in the romantic writer Coleridge's opium-induced poem "Kubla Khan." Nelson's style and approach make heavy use of romantic tropes. From the emphasis on truth discovered in personal exploration, to the celebration of dreams, visions, and revolution, to the suspicion of technical rationality, and to the strategic use of vernacular language, Nelson has crafted his own version of orthodox romantic style. Nelson's enthusiasm for technology certainly distinguishes him from, say, Thoreau or Wordsworth (who famously wrote in "Tintern Abbey," "For nature then/To me was all in all.") But the original romantics were never opposed to technological advances in the same fashion as, say, the Amish.[52] The original romantics were products of the emerging new technological world; they raised questions about that world and pointed to what they saw as its limits and spiritual failings, but they were not really ones to step completely outside of it. They lived and moved about in the new world being created by new technologies of communication and transportation, regularly riding the railroad into the countryside, living off of an economy made possible by the rotary printing press. And they were not beyond considering the artistic character of new technologies. In his 1833 sonnet, "Steam-boats, Viaducts, and Railways," Wordsworth allows that these machines should be embraced by "Nature" because they are products of "Man's art." More enthusiastically, Walt Whitman wrote an eroticized celebration of the locomotive, "To a Locomotive in Winter" ("Thy metrical, now swelling pant and roar—now tapering in the distance/Thy great protruding head-light, fix'd in front/Thy long, pale, floating vapor-pennants, tinged with delicate purple").[53]

But the real thread that ties Nelson to traditional romantic writing is the creation or expression of a distinct, struggling persona. This is not just an idea or a narrative but a collection of textual practices that construct a very particular relation between writer and reader. Robert Darnton has suggested that the origins of certain modern patterns of reading and writing can be traced back to the time of Jean-Jacques Rousseau.[54] Rousseau, Darnton has argued, did much to "fabricate romantic sensitivity" by "transforming the relation between writer and reader, between reader and text." At the core of this new rhetorical situation was an effort to put the persona of the writer in the forefront. "Instead of hiding behind the narrative and pulling strings to manipulate the characters in the manner of Voltaire," Darnton writes, "Rousseau threw himself into his works and expected the reader to do the same." Rousseau encouraged his readers to approach his works as the authentic, unmediated expression of the inner feelings of a unique human being. Rousseau envisioned a form of art that was capable of "communicating to those far away, without any mediation, our feelings, will, desires." In *La Nouvelle Heloise*, for example, Rousseau, not only made the then-unusual gesture of signing his own name to the novel, but also made much of that fact in the preface, insisting that a "man of integrity" should not hide himself from the public. He furthermore insisted that "I do not want to be considered any better than I am."[55] "A man of integrity", in other words, is a man who bares his flaws, which in turn become the mark of authenticity, the sign of an honest connection.

This now-familiar understanding of the nature of writing and reading—Promethean authors sharing their inner struggles with their readers—echoes in many arenas beyond the expected ones like secondary school literature classrooms. They are not unheard-of in academic writing, for example. Rousseauian textual constructs were originally contrasted with what Rousseau took as the stale, inauthentic, contrived writing of the Parisian salon; there are echoes of this aspect of Rousseau in common criticisms of academic jargon and modishness, and to ameliorate such criticisms it is often fashionable for us academics to adopt some post-Rousseauian tricks in our otherwise salonlike writing. For example, we cultivate a few elements of a unique writing voice or strategically insert a personal detail or two into our treatises.

There are many variations of the romantic rebel hero, from Thoreau to Van Gogh to Che Guevera. But if Nelson has a predecessor, it is probably William Blake, the late eighteenth-century English poet. They are not particularly similar as intellects—Blake was a religious mystic and critic of industrialism, whereas Nelson is more of a libertarian—and there is no evidence that Nelson was particularly influenced by Blake. But the point of comparing them is that they both offered a similar reading experience and resonated with their audiences by way of a similar set of textual practices. An analogy with Blake helps explain the particularly compelling nature of Nelson's writing, the *effect* of Nelson's oeuvre on readers of the 1970s and 1980s.

Born into a working-class family, Blake was trained as an engraver. Engraving was the primary method of mass producing images in the eighteenth century, and thus in a sense it was the multimedia of the day. When Blake sought to make his own art, he combined poetry, drawing, and mechanical reproduction to create a medium that was appropriate to his intense personal vision; he invented a new form of copper-plate engraving and eschewed the industrialized uniformity of moveable type. He was thus able to create a unique form of illuminated manuscript, with the text as part of the engraving and each page hand-colored. Like Ted Nelson, Blake was more committed to his vision and his persona as a rebel than to economic or professional success; Blake's insistence on maintaining the integrity of his work by eschewing traditional printing techniques ensured that he would never become widely known or successful during his own lifetime.

Also, like Nelson, Blake frequently presented his personal philosophy in witty, biting aphorisms and was known for constantly coining new terms. Blake's "Proverbs of Hell" are some of his most widely quoted material. For example, "All that is now proved, was once only imagined," attacks empiricism. "Without contraries is no progress," attacks rationalist deductivism. And "Prisons are built with stones of Law, Brothels with bricks of Religion" expresses Blake's views of conventional morality. Nelson is similarly famous for the pithy statement: "Computers are no more inhuman than we make them." "Computers are wind-up crossword puzzles." Blake's longer works were based on his personal mythopoetic universe, replete with his own eccentrically concocted mythic characters. Nelson has famously coined a universe of terms, of which hypertext was merely the beginning; for example, "transclusion" (for a system that allows the inclusion of portions of other documents in newer, linked documents); "transcopyright" (Nelson's vision of an automated system of micropayments for reading and linking to others' works); "thinkertoys" (for open-ended devices intended to enable intellectual exploration and experimentation); and "intertwingularity" (for the nonhierarchical, interrelated form of most knowledge).

Nelson, in sum, can be described as a libertarian, computerphilic counterpart to Blake; both are witty, aphoristic, philosophically sweeping, economically inauspicious writers and self-described radicals who do their own illustrations and are given to endless neologisms. And both are captivated with the idea of unmediated, personal control over the process of producing texts for the purpose of self-expression—in Blake's case, via hand-colored etchings, in Nelson's, via networked computers.

How and why did all this matter in the development of our computer-networked world? The point here is not to draw psychological parallels between the two thinkers but to point to parallels in how and why they are *read*. There is a particular kind of reading pleasure offered by an immersion in the works of Blake or Nelson, and this in turn helps explain the impact of such works on readership.

Both Blake and Nelson offer precisely an intimate encounter with a *persona*, a specific literary interaction with a constructed unique individual. Beyond a few of his most frequently anthologized poems, Blake's appeal is inextricable from the reader's developing a sense of Blake as a person. In reading Blake, one not only encounters his startling insights and compelling views; in learning to adjust to his eccentric spellings, his mythopoetic characters and terms, to his drawing and coloring style—which is less beautiful than compelling and didactic—one becomes accustomed to him the same way one might develop a fondness over time for the quirks of a loved one. The pleasure of slogging through his often dense and eccentric works is bound up with the enjoyment of reading his persona—not in the sense of getting to know the details of his life or his times, but in the sense that what one is reading is unique to the intellectual process of a particular author. It is not that his eccentricities are things to be overcome in order to get the universals in his works; on the contrary, those eccentricities are part of how the texts convey a sense of the individual person of Blake embarked on a process of discovery.

Reading Ted Nelson's *Computer Lib* is similar. While browsing the often prescient and arresting insights into institutions and technologies, one also develops a familiarity with Nelson's handwriting (most of the titles of sections are hand written), his hand-drawn illustrations (often in comic strip format), and, of course, his distinct writing style. And one gains a sympathy for his own pleasures in computing and personal frustrations with the problems of the field. For both Blake and Nelson, one sometimes feels awe at their determination, rebelliousness, and accomplishment, sometimes an appreciation for their insights and values, and sometimes a kind of poignant identification with someone who gives the impression, not so much of choosing not to compromise with the dominant ways of his world, but of someone who could not do so. And of course part of the pleasure is one of identification with the rebel hero; even if only in a fleeting way, there's a thrill in imagining that we readers, too, might know more than the authorities, that we might be right when our bosses are wrong.

By the late 1970s, *Computer Lib* and other writings by Ted Nelson were becoming familiar and, in significant numbers of cases beloved, among various pockets of the invisible colleges of individuals who developed and programmed computers.

Why a *Personal* Computer?

In the 1960s, most of the innovations that went into the use of computers as communications devices—graphic interfaces, email, discussion lists, user-friendliness in general—were developed by individuals lower down on engineering hierarchies, individuals whose salaries were paid for by projects officially dedicated to other purposes, like fighting nuclear wars, connecting research scientists to super computers for elaborate calculations, or Taylorizing routine office work. In the 1970s, the same could be said for the microcomputer; it appeared in the margins of the industry.

The term *personal computer* crept into the language in the mid-1970s, quickly becoming attached to the early hobbyist computers from the Altair onward, and became enshrined in the abbreviation PC where it remains in the language today. Other terms have coexisted with it—*micro-, desktop-, home-*—but the vaguer *personal* seems to have endured. Why? The word *personal* as an adjective for a gadget does not self-evidently mean it is designed for use by one person. We do not call watches "personal clocks," transistor radios "personal radios," pocket calculators "personal calculators," or cell phones "personal telephones."

The word *personal* entered the vocabulary of computing because it is the opposite of *impersonal*. Before the mid-1970s, both the computer industry and the culture at large generally saw computers as the embodiment of the neutral, the universal, the rational and mathematical—as impersonal, as tools for centralizing bureaucracies, Taylorizing the office, or winning nuclear wars (or, like *2001's* forbidding HAL, as potentially murderous artificial minds). Like the slogan "black is beautiful" in the 1960s, "personal computer" was a deliberate combination of two things the dominant culture understood as opposites. At the beginning, attaching the term *personal* to something associated with impersonal universality provided a nicely startling juxtaposition, a two-word condensation of a larger cultural refiguration of the meaning of computing as a whole. It announced a radical reclassification of computers, taking them out of the old box of mathematical impersonality and putting them in a new one that associated them precisely with individual uniqueness, distinctiveness, unpredictability, and expression—with all those things we have long associated with the romantic persona.

Such a refiguration needs a larger context. By 1975, Nelson's *Computer Lib* and the alternative ethos of computing it advocated spread into pockets of the larger community of those with some interest and expertise in computing. The cover of the September 1976 issue of *Byte* magazine sported a playful, hand-drawn image of a sixties-style political rally, with attendees holding aloft signs that say "Two Computers in Every Home!" "Computer Power," and—quoting Nelson— "Stamp Out Cyber-Crud!" One of the people in the crowd wears a T-shirt emblazoned with the cover image from Nelson's *Computer Lib*: a raised fist. And the image contains a now-familiar sense of computer-culture whimsy: *Star Trek's* Spock is in the crowd, and a starship Enterprise flies overhead.[56] It would be several years before this kind of imagery would be associated with computers in the broad popular imagination (HAL was still the more common image of computers in the pop culture of the mid-1970s), but to those inside the various invisible colleges associated with computer engineering, these images were becoming familiar.

This is the period in which computer hobbyists and tinkerers invented the microcomputer and revolutionized the structure and character of the industry. The first popular hobbyist computers appeared for sale in 1976, and within a decade new industrial empires would be born, old ones would be in a state of crisis, and the entire industry would look radically different. The core events have been much mythologized elsewhere in books, documentaries, and even docudramas.[57] Basically, in the mid-1970s, off the radar of the major corporate and military players in the industry, communities of computer hobbyists began tinkering with the ever-cheaper digital microchips, most famously the attendees at the Homebrew Computer Club in Palo Alto. Organized by political activist Lee Felsenstein, who wanted to bring computing to the people in true 1960s fashion, the club members freely shared information with one another, including things like the version of the Basic computer programming language that had been written by college dropout Bill Gates for the Altair, the first commercially successful hobbyist computer. Steve Jobs and Steve Wozniak were regular contributors and built the Apple I to impress their friends in the club.

The rise of the microcomputer in the 1970s is worth briefly discussing in the context of the origins of the internet for two reasons. The first is context. Microcomputers both spread the experience of interactive computing and created a context for widespread networking that would begin roughly fifteen years later. Second, the 1970s microcomputer revolution nicely illustrates the complexity of the relation between cultural trends and technological developments.

It would be an oversimplification to draw a straight line from Nelson's *Computer Lib* to the appearance of microcomputers in the late 1970s. Yes, Nelson was known to many computer hobbyists of the time and made appearances at the Palo Alto Homebrew Computer Club specifically. But the first generation of microcomputers that emerged from the era, from the Altair to the Apple II to the IBM PC, were hardly the graphics-intensive, intuitive "dream machines" advocated by Nelson. (Nelson himself complained that the Homebrew crowd was too obsessed with gadgetry, with chips and wiring, at the expense of elegant software design.)[58] Moreover, one cannot reasonably argue that Ted Nelson's work was *the* cause of the eventual triumph of user-friendly, graphics-oriented computing; there were others in the 1970s and 1980s who were promulgating various flavors of the Engelbartian approach to computing, such as Andries Van Dam at Brown, Xerox PARC, and so on. The revolution, in some form, would likely have happened without him.

To search for which individual originated which innovation, however, is to misunderstand the character of sociotechnical change. By the mid-1970s, most of the possible notions about how computers might be used and understood had already been articulated, and the microchip industry was already well on its way towards making computing inexpensive; Moore's Law had already become an operating principle. The real question is not who invented what but how certain visions and uses became institutionally embedded and enacted. If Ted Nelson and the countercultural articulation of computing made a material difference in the world, it was on the level of changing how individuals understood themselves within institutions. Specifically, it offered a different self-concept, a different picture of who one was when using and building a computer.

By the mid-1970s, several different visions of computing were at play in the technical community. While the corporate community was struggling with the floundering effort to implement Taylorized "offices of the future," the military was imagining global command-and-control systems with the ARPANET, and descendants of Engelbart were exploring the encyclopedic vision of computing, Ted Nelson was the most important spokesperson of a community promulgating a distinctly countercultural vision of computers as creative writing machines that enabled self-exploration and expression.

To understand the impact of these competing discourses or visions, it helps to remember that the microcomputers these hobbyists were building were not all that unique in a purely technical sense. In fact, the first microcomputers were not technologically all that different from the machines being sold by the big manufacturers to implement the "office of the future." For example, in 1977—the same year that Apple began selling the Apple II computer—IBM introduced its System 6 "information processor," which was described at the time as "a terminal with a small TV-like screen to display text, a 'floppy disk' memory that stores more information, and a high-speed inkjet printer that controls a flow of ink droplets to form characters on paper [and which could] communicate with a computer or with other IBM word processors over phone lines."[59] The sticker price for an

Apple II was much less—about $2000[60] compared to $16,450 for the IBM system—but the latter included a floppy drive, a monitor, networking capabilities, an ink-jet printer, and elaborate software. These items, if they could have been added to the Apple II at the time of its introduction, could well have brought it to a similar price range.[61]

What distinguished the Apple II from IBM's System 6, then, was less the specific technology than the imagined use. One bought an IBM System 6 in the expectation of solving particular problems already institutionally defined, the kind of problem that could be laid out in a grant application or a corporate business plan. One bought an Apple II in 1977 simply to have a computer, simply to see what it could do, to *explore*, not to undertake a known task. The point of view embedded in System 6—in its marketing, programming, cost structure, and so forth—was that of upper management concerned about cutting costs and better regulating behavior within giant, far-flung enterprises. The Apple II was designed with another ethos in mind, purposes that in turn implied different views about how the social world worked. The Apple II was much cheaper than the IBM System 6, not just because of Steve Wozniak's famously clever circuit design, but just as importantly because it did not come as a complete system ready for integration into a corporate office; it was sold simply as a box that could be plugged into a monitor or a TV set, without a printer, disk drive, or elaborate set of software dedicated to corporate goals. The Apple II was remarkably cheap only if one's goal was to *play with a computer*, only if the sheer fact of owning and operating a computer was a goal in and of itself. IBM's System 6 and similar machines were created in a context where such a goal conflicted with the basic instrumental understanding of what a computer was for. In an important sense, the microcomputer was not a new technology; it was a new way of imagining, marketing, and using existing technologies. An Apple II was supposed to offer suprises, whereas a System 6 was supposed to prevent them.

In the late 1970s, computer fairs and other industry conventions were an important locus for cultivating shared interpretive frameworks for the industry. Engineers, executives, gadgets and reporters were brought together in physical proximity. This would allow, not just the airing of new ideas, but that rich, exciting sense of affirmation and amplification that comes from being face to face with others who share one's view. In June 1977, the *New York Times* sent a reporter to the National Computer Conference in Dallas, whose story prominently featured the new excitement around microcomputers.[62] In August, in an article specifically about the enthusiasm around microcomputers, the paper called readers' attention to the upcoming Personal Computing Show in Atlantic City, Computermania in Boston, and the Personal Computing Expo in New York City. And the article referred back to the now-legendary debut of the West Coast Computer Faire in

San Francisco, which took place in April.[63] Within an hour's drive of Silicon Valley, the West Coast Computer Faire offered attendees the first look at the Apple II, attendance was double that anticipated,[64] and it was becoming clear to those in attendance that the new world of microcomputers was going to have an impact. Nelson gave a keynote speech called "Those Unforgettable Next Two Years," and he opened by saying, "Here we are at the brink of a new world. Small computers are going to remake our society, and you know it." He continued,

> For now, though, the dinky computers are working magic enough. They will bring about changes in society as radical those brought about by the telephone or the automobile. The little computers are here, you can buy them on your plastic charge card, and the available accessories include disc storage, graphic displays, interactive games, programmable turtles that draw pictures on butcher paper, and goodness knows what else. Here we have all the makings of a fad, it is fast blossoming into a cult, and soon it will mature in to a full-blown consumer market. . . . The rush will be on. The American manufacturing publicity machine will go out of its gourd. And the next two years will be unforgettable.[65]

It would be five years before the microcomputer would be featured on the cover of *Time* as "man of the year," not two,[66] but again Nelson's predictions would turn out to be surprisingly accurate, more so than most of the musings about the future of computing that were appearing in the mainstream press at the same time, which still tended towards visions of the Taylorized office. As a result, thousands of individuals who were involved or were becoming involved with computing by attending conferences (or talking to those who attended them) were exposed to the countercultural rendition of the meaning of computing, and over the next several years would experience first hand how, on occasion, a rollicking iconoclastic discourse could turn out to be more accurate than the dry, jargon-ridden prognostications of mainstream executives, academics, and financial page reporters.

Conclusion

At the end of the 1970s, the various visions of computing associated with Ted Nelson— computers were for symbol manipulation and therefore they should be used as vehicles for passionate exploration and self-expression—were still minority views, both inside the industry and out. The big money was still flowing towards giant mainframes, centralizing corporate applications, military applications, and exotic experiments like artificial intelligence—approaches to computing that, for all their variability, were hardly countercultural. Even those who were interested in computer communication tended to think in rationalist, Enlightenment terms; computers were going to be for sober activities like looking up scientific information, helping professionals keep their appointments, or better educating youth.

During the 1970s, however, a new cultural toolkit was made available to and rendered compelling within the world of computing. Its significance was not just that it brought bean bag chairs and other countercultural trappings into the office buildings of the high-tech industry. Rather, it offered a new social meaning for computer use, a new vision of what it meant to sit down at a computer console and of who the person was who was using it—a new idea of self in association with computers. It was not news to computer engineers of the 1970s that in some sense computers could be fun, that there existed computer games, that computers had a holding power. But, before the counterculture, that knowledge and experience for the most part had to be treated as an insider secret or an odd side-effect of the machines that one mentioned guiltily, if at all. Computers needed to perform specific, rational functions for large organizations; that they might also be fun was not something to put in a business plan, grant proposal, or marketing campaign—not something that might help legitimate computers.

By the second half of the 1970s, that was changing. From about 1976 onwards, it was common enough for a computer engineer or graduate student, perhaps one whose livelihood was secured by money from the Pentagon, to pick up a copy of *Computer Lib* or *Creative Computing* during a coffee break and, in the midst of descriptions of new programming languages or techniques for rendering graphics, be exposed to a mode of understanding, to what Bourdieu calls a *habitus*, in which the compulsive character of computer work might be associated with social legitimacy, might be something that could be brought into the light of public advocacy. By offering a romantic framing of computer use—computer use could be articulated as playful, expressive, even rebellious—the activity of computer use and design no longer need be instrumentally tied to a specific ends; the means could be an end in itself.

Most of the effects of this romantic framing, as we will see, would come in subsequent decades; in 1980, the broad outlines of the computer industry and its place in the larger culture was not all that different from the situation of 1970. But the computer counterculture's survival was secured by microcomputers, which were originally conceived and sold to a degree, not for specific purposes, but in a playful, computing-for-computing's sake way. This not only created a new market but a widely accessible alternative vision to the highly ends-oriented, instrumental way in which computers had been imagined up to that point. This fact left its imprint on the machines themselves; the original entry-level IBM PC would offer a game port before it offered a hard disk.[67] But, perhaps more significantly, this playfulness of means and uncertainty of ends was now potentially connected with a rebel-hero identity. That identity would provide people in the profession with a new way of thinking of themselves and their relations to others and would also draw new people into the profession.

3 Missing the Net

The 1980s, Microcomputers, and the Rise of Neoliberalism

They showed me . . . a networked computer system. . . . They had over a hundred Alto computers all networked using email etc., etc., I didn't even see that. I was so blinded by the first thing they showed me which was the graphical user interface.[1]

—Steve Jobs, looking back on his visit to
Xerox PARC in December 1979

Two Guys in a Garage?

The Apple II computer was initially the product of the collaboration of three people, Steven Jobs, Steven Wozniak, and Armas Clifford "Mike" Markkula, Jr. Markkula, a Silicon Valley venture capitalist, was clearly essential to the creation and success of the company; Wozniak has suggested Markkula was the most important of the three.[2] He provided both capital and managerial skills, served as chairman of the board, and for a while during Apple's period of most rapid growth, he was Apple's CEO.[3] The Apple was not the first microcomputer; there were many hobbyists tinkering with tiny computers at the time Apple was started, and already some of them were manufacturing and selling them. What distinguished Apple is that it led the fledgling industry beyond the hobbyist market into the larger world. Markkula, arguably, is the one who made this happen, who distinguished Apple from all the other early microcomputer builders by using his knowledge and connections to turn a business run by and for hobbyists into something capable of growth beyond those bounds.

Yet a search of English language newspapers and magazines for the decade of the 1980s turns up only 83 articles that mention Markkula. Steve Wozniak turns up 417 articles, and Steve Jobs, 791, a difference of five-fold and ten-fold, respectively.[4] Apple computer, the world was repeatedly told, was started by "two guys in a garage," and Markkula was not one of those two guys.

Why this oversight? In 1985, newly reelected President Ronald Reagan announced, "We have lived through the age of big industry and the age of the giant corporation. But I believe that this is the age of the entrepreneur."[5] This was Markkula's problem. He did not neatly fit the mythic American narrative of the entrepreneur, who in popular fantasy came from nowhere and needed no outside support. Markkula came from somewhere and had connections. He was an experienced Silicon Valley manager, fluent with the complexities of incorporation, venture capital, manufacture, and distribution. He brought an established body of knowledge, social relationships, and experience to bear on the production and marketing of the microcomputer. But, in the 1980s, the mainstream press was sensitized to the story of the entrepreneur and was eagerly looking for real-world instances of the narrative. And entrepreneurs, the classic narrative goes, work alone, without connections, background knowledge, or established social frameworks. A search in Lexis/Nexis for articles that contain the words *garage* and *Apple Computer* in the 1980s turns up dozens of articles with titles like "More Young Millionaires, Please," (*The Economist*),[6] "Risk Takers" (*US News and World Report*),[7] and "The Spirit of Independence" (*Inc.*).[8]

It is a fact that, in the year before Markkula arrived, Jobs and Wozniak, working literally out of a garage, did make and sell about two hundred circuit boards that could form the core of a hobbyist computer called the Apple. This nugget then became the core of the hugely popular entrepreneurial fable that Apple computer was started by two guys in a garage. But this pre-Markkula year involved an unincorporated, relatively informal partnership of two college-aged hobbyists making something for other hobbyists. Wozniak was working full time for Hewlett Packard at the time and has said of the change when they joined up with Markkula, "This was different than the year we spent throwing the Apple I together in the garage. This was a real *company*. I designed a computer because I like to design, to show off at the club. My motivation was not to have a company and make money."[9] But, in the 1980s, those things that distinguish two guys in a garage from a real company would largely disappear in media coverage of the microcomputer industry, casualties of the entrepreneurial narrative.

This lacuna was a symptom of a general transformation in the dominant, governing ideas of American society in the early 1980s, when a radical belief in markets and an accompanying suspicion of all forms of government regulation—beliefs that were once thought to be fringe—would become common sense among many in positions of power, with global effects. In the early 1980s, too, the image of computers as distant and formidable was overthrown in the popular imagination and replaced by the little typewriter-like boxes appearing with ever-more frequency in advertisements, the media, offices, and homes. These two epochal shifts were not unrelated. The particular way the microcomputer appeared on the American scene, we shall see, helped make the market faith *feel* right. At the same time, the market lens distracted the larger culture from developments that would later prove to be momentous: specifically, developments in the internet.

Microcomputers and Markets

John Maynard Keynes famously said, "Practical men, who believe themselves to be quite exempt from any intellectual influence, are usually the slaves of some defunct economist."[10] True enough, but why sometimes Marx and other times Milton Friedman? What makes some intellectual frameworks gain widespread support while others disappear? Sometimes there are broad forces at work, like economic self-interest. But there are too many cases of wealthy Marxists and working-class economic conservatives for that to be the whole story.

Of course, part of what drives the uptake of ideas is their careful cultivation in traditional institutions, like academia, think tanks, and similar institutions, followed by the careful promotion of ideas into various halls of power through publications, conferences, and the money to produce them; in the case of neoliberalism, we will see that the Chicago law and economics movement and a series of publications about the "information society" played their part. Sometimes that is all that is needed; numerous ideas filter from institutions into our businesses, laws, and legislation without the general public hearing much about them, much less understanding them.

But, on a broader more enduring level, there's another piece of the puzzle. To really influence society widely and deeply, ideas need to become vivid. Connections need to be drawn between the structures of everyday lives and the larger, more abstract world. For example, rural life offers experiences that lend themselves to certain political worldviews—a hunter's distaste for the annoyances of gun licensing might evolve into a distaste for government regulation in general. And life in a city offers different experiences, in which a concern about urban violence or zoning laws might make one more receptive to regulation in general. Everyday life, however, is not just a geographical place like suburbia. It is built out of the regular engagement with people, spaces, and objects.

In the 1980s, for a large chunk of the U.S. middle class, the microcomputer became just such a socially evocative object in their everyday lives, an object that brought with it certain experiences from which one might draw broader conclusions about the nature of social existence. Specifically, the way in which microcomputers appeared in everyday life helped turn them into an emblem of what's good about the free market.

Culturally, two things were significant about this period. First, networking was ignored because the dominant culture was seeing things through free market lenses and thus imagined that microcomputers were about isolated individuals acting on their own; it was disinclined to think about the broader social relations like networking that both produced microcomputers and that could be enabled by them. The focus of this chapter, then, will be on what preoccupied the dominant technological imagination of the time, particularly how the microcomputer played a role in the formation of what has since become known as neoliberalism or, to its critics, market fundamentalism. Second—and this will be the subject of chapter four—in the crevices of the military-industrial complex, a specific set of practices evolved associated with early networks that in fact pointed in a very different political economic direction. The community of computer networkers relied on a high degree of awareness of the complexities of social relations that everyone else was ignoring; enriched by elements of countercultural style, this community was inventing the very nonmarket tradition of open software production via "rough consensus and working code," a tradition that would lead to the surprising rise of the internet in the early 1990s and later in the decade become the core of one of the major countervailing forces against neoliberalism's simplistic market vision.

How to Make Markets Modern:
Technology and the Rise of Neoliberalism

In legal thought, political rhetoric, and in popular culture, the United States has always had strong strains of a Lockean individualist framing of life, wherein the power of individuals to pursue profit in a marketplace is understood as the archetype of freedom. But the U.S. has never been purely a place of *Homo economicus*. For example, patriotism, religion, ethnocentrism, civic republicanism, Emersonian individualism, the labor movement, and the sentimentalized middle-class notion of the home and family have all at points worked as popular and powerful counterweights to strict marketplace individualism. Various flavors of socialism and anarchism, furthermore, have their important place in American traditions. Market individualism has been not so much *the* American ideology as it has been a regular part of the mix of ideologies that, at any given moment, comprise the social and political terrain.

It was in that sense that, beginning in the late 1970s, a new variety of market individualism arose and came to have enormous influence on policy making both nationally and internationally from about 1980 through the end of the twentieth century. Known variously as neoliberalism, the Washington consensus, and—derogatively—as market fundamentalism, as of this writing it has only recently lost its overwhelming dominance of policy making in the United States and most other parts of the world and still remains a powerful force.

Intellectual movements can thrive inside universities and other institutions without ever gaining traction in the worlds of politics and popular culture. But any movement that does gain acceptance in politics and culture needs some form of expert institutional legitimacy and cultivation to succeed. Neoliberalism triumphed because of its articulation with popular consciousness, but first it gained a platform inside various think tanks and academic movements. Technology played no small role in each stage of its cultivation.

Technology in the Law and Economics Movement

There was a time when new technology and progress seemed to belong more to the left side of American politics. In the Progressive Era, Louis Brandeis was able to argue the case for antitrust law by painting trusts and robber barons as unscientific and thus inefficient and against progress.[11] In the 1930s, New Dealers celebrated the government-created TVA and other big government projects as being on the side of new technology and progress.[12] The fact is, the first half of the twentieth century saw the rise of giant corporations like AT&T and General Electric, the transformative effects of the New Deal, and the centrally managed war effort of the 1940s, all of which provided the context in which the United States had become the richest and most powerful country in the world and its citizenry had been showered with wondrous new things like radio, refrigerators, television, automobiles, and the interstate highway system. And much of this had been accomplished, not by plucky entrepreneurs striking out on their own, but by relatively tight cooperation between government and large, multiunit, bureaucratically centralized corporations. Keynesianism gradually became orthodoxy for both political parties, powerful intellectual leaders wrote books like Eugene V. Rostow's 1959 *Planning for Freedom*, and modest government regulation of everything from television news to airline ticket prices seemed rational, professional, and forward-looking.[13]

By the 1960s, as a result, American advocates of orthodox economic libertarian principles looked backward and old-fashioned. Calls for dramatic rollbacks in government regulation and praise for the competitive free market seemed throwbacks to the previous century, as quaint and out-of-date as *Robinson Crusoe*.

What the economic conservatives astutely recognized was that, to regain intellectual authority, they would need to make market individualism modern. The works of Ayn Rand, Hayek, and some marginalist economic theory might be persuasive in their own circles of true believers, but the larger world needed something more. So they set out to show, not just that markets in general were efficient or moral, but more specifically that free markets, unhindered by government regulation, could better handle the most modern of technologies. Radio, television, jet planes, even computers, the underlying argument went, did not need government regulation like the FCC and Federal Aviation Administration, or government-funded research, or protected, regulated monopoly corporations like AT&T; on the contrary, they needed to be freed of the shackles of all these things. Free markets could be high tech.

Several different strains of thought combined to provide an intellectual framework for all this. Most famously, a fully deliberate effort centered at the University of Chicago's *Journal of Law and Economics* grafted strains of neoclassical economic thought onto key legal concepts, and other neoclassically oriented economists began seeking ways to give new life to the idea that government regulation was an anathema.[14] Beginning in the 1960s, a series of up-and-coming conservative activist intellectuals, such as Richard Posner, Douglas H. Ginsburg, and Robert Bork, sometimes in association with foundations like the classically conservative Heritage Foundation or the more libertarian Cato Institute, began publishing articles that probed the intellectual apparatus of the twentieth-century American welfare state for weak spots. Not content to stick to traditional, conservative fables about village markets, Horatio Algers, or celebrations of corporate chieftains, they tackled the *hard cases*: airline regulations, antitrust law, FCC regulation of radio waves—exactly those cases that New Dealers and their successors had previously identified to illustrate the limits of free markets. In each case, the law and economics scholars tried to show that the cozy relationships between business and government characteristic of postwar corporate liberalism—relationships like those encouraged by the reasoning of Vannevar Bush's *Science, the Endless Frontier*—were, despite all appearances, neither necessary nor efficient and would be best dealt with by dramatically scaling back government money and regulation. Lawyers, judges, and politicians, the argument went, could make sound, forward-thinking decisions by thinking in terms of notions like consumer welfare and economic efficiency instead of concepts like the public good.

The law and economics activists, it's worth remembering, were offering what seemed like bright new ideas in the context of a vacuum. The early 1970s, with the Watergate scandal and the defeat in Vietnam, is often read as a low point for American conservatives. But it was in fact a low point for a certain kind of postwar, nationalist, corporate-oriented centrism; the Nixon administration had annoyed both social and economic conservatives with many of its policies, and the Vietnam debacle had been set in motion by Democratic presidents. And, as the 1970s progressed, the U.S. economy was struggling and industrial leadership was bewildered; these were the days of the short-lived Ford and Carter administrations, of stagflation, and of the initial stages of the devastating decline in rust belt manufacturing. In this depressing context, the law and economics theorists offered, not only criticisms, but what to some looked like a way out of this morass, one that would not seem to threaten existing wealth and ways of life.

Information Society Theory

Roughly contemporaneously, a rather different strain of thought relevant specifically to computers appeared, generated by the information society theorists. In the leadership vacuum of the mid-1970s, the rhetoric of the information society provided another galvanizing alternative vision for capitalist energies.

In the ashes of the 1960s counterculture and McLuhanism, Daniel Bell, Marc Porat, and others began to focus on the centrality and importance of various symbolic economies in developed industrial societies; corporations, governments, and everyday life seemed ever more dependent upon and infiltrated with ever more elaborate systems of communication, like satellites and computers, and ever more elaborate forms of data and cultural products, like marketing research, globally distributed Hollywood movies, and so on.[15] Jean-Francois Lyotard noticed these trends around 1980 and wrote *The Postmodern Condition*, sending a faction of humanists off on a two-decades intellectual romp called postmodernism.[16] But, more in tune with the power centers of Washington, DC, Bell and Porat interpreted these trends as signs of a coming information society, where, as Bell put it, life would no longer be about "the management of things" but instead "a game between people."[17] The refrain that emerged from these texts was that we were approaching a society where the principal commodities would not be traditional resources or industrial objects but rather digitally distributed information.

Numerous compelling scholarly critiques of the information society tradition have been published. Much of the early work, it has been argued, rested on a misinterpretation of the reports of a growing service sector of the economy in which the growth of low wage service jobs like restaurant work is mistaken for a growth in knowledge work like computer programming. Also, the argument for a fundamental transformation in capitalism (as opposed to a simple continued development in long-standing trends like consumerism and the use of telecommunications to coordinate production) was thin, more often assumed than demonstrated.[18] Among the sociologists and economists who make a living seriously studying epochal historical change, few have taken the idea of an information society all that seriously.[19]

But the idea of the information society nonetheless had an enormous appeal outside academe, most evidenced by its popularization in works of Alvin Toffler, who added to his line of future shock bestsellers with the 1980 *Third Wave*. So why did this notion do as well as it did? It could indeed be readily foreseen that communication systems were on a trajectory towards convergence in the digital. But there were a lot of other terms referring to the same phenomena that were thrown out at the time—*technetronic society, telematic society, compunications*.[20] All of these terms were based on the expectation of a gradual convergence of media enabled by digitalization, and all were to various degrees embedded in versions of what Carey and Quirk called the "rhetoric of the electrical sublime." They all shared the same technological determinism, industrial optimism, and quasi-religious sense of progress and transcendence through technology. And they tended to share the low threshold of plausibility that Tom Wolfe pointed out was central to McLuhanism in his essay called "What If He Is Right?" It wasn't that you had to be fully persuaded that television was creating a global village or that we were leaving the industrial age and entering an information one; all you had to think was "what if he's right?" and you'd pay attention.[21]

But most of the terms competing for the same discursive space focused on the role of networks and computer technologies, on the machinery. The term *information society* had a special appeal that caused it to outlast all these other buzz words and to periodically resurface as it did in the early 1990s with the rhetoric of the information superhighway.

The word *information* suggests that meaning can be treated as a thing and thus as manageable. As intellectuals looked towards the eventual digitalization and convergence of media, they could see that walls were breaking down and collapsing into something dispersed and granular; but among the ambitious capitalist elite that something was not Baudrillard's implosion of meaning or Lyotard's end of totalizing discourses but rather the rise of information. From the point of view of the power structure of capitalism, information had the extraordinary advantage of being something you could imagine as thinglike and therefore as property, as something capable of being bought and sold. And this had a broad appeal to a struggling corporate leadership. If you said we were moving into a telematic society, you might attract the attention of manufacturers of networking and telecommunications equipment, but if you said we were moving into an information society, the appeal was wider, ranging from Hollywood executives to university administrators to Wall Street brokers.

Synthesis: Digital Convergence and the Construction of Digital Information as Property

In the narrow but powerful worlds of think tanks, Federal administrative agencies, Congressional staffs, and corporate government relations offices, information society rhetoric and neoclassical economics intersected, with powerful effects. A useful illustration of how the information society rhetoric and neoclassical economics all came together can be found in the realm of intellectual property. In the 1970s, the idea that software could or should be owned, particularly on the level of algorithms and patents, was at best controversial. The inventors of the spreadsheet did not patent the concept because in the late 1970s that kind of thing wasn't considered patentable. A few years later, however, Bill Gates parlayed patented control over MSDOS software into what would become one of the world's wealthiest corporations.

This change in thinking in the courts and legislatures was made possible by a confluence of ideas. Significantly, the discourse of the information society *assumed* that information already was a known and quantifiable thing. The information society pundits did not say we should turn information into property; they said it *is* a thing, it *is* data on hard disks and thus inherently property (or at least is easily turned into a commodity) and derived their analyses from what they took to be that self-evident fact. (This is why the pundits of the day largely failed to foresee the Napster problem, that is, the way that computer communications would disperse and accelerate the process of copying texts so as to problematize the whole notion of property; they thought that the treatment of information as property would emerge self-evidently from the technology, and so it rarely occurred to them that in fact these very technologies might undermine the idea of property.) They assumed that, since to their minds digital information simply *was* property, not something that would have to be awkwardly shoehorned into the framework of property, the digitalization of culture would proceed perfunctorily.

This was then coupled to the new enthusiasm for open markets. This coupling was not driven exactly by logic. In the 1970s, the scholarly law and economics conservatives spent most of their energies on things like broadcast spectrum and airline deregulation. They did not focus as heavily on intellectual property at the time, perhaps because, from the point of view of their theory, the justification for copyrights and patents are thin. Copyrights and patents create temporary monopolies and can encourage what economists might call inefficient and anticompetitive rent-seeking behaviors. If neoclassical economics doesn't exactly encourage intellectual property, however, *rights discourse* does. The law and economics movement did help revive the language of classical property rights and this "rights talk" lent itself in a general way to the notion that the more rights the better.

What happened, then, was a loose fusion of information society rhetoric with the general law and economics enthusiasm for marketization within the courts; this fusion laid the foundation for the early 1980s legal decisions that began to inflate the notion of intellectual property. Key trends at the time were the extension of patents to genes and software, and later intellectual property protections extended to the look and feel of a computer program, to biological cultures and genetic sequences, to many aspects of a pop star's personality, and eventually in the 1990s to things like business models and Amazon's one-click book ordering. Beginning in 1984, the U.S. Congress and the U.S. Department of State got in the habit of demanding that other nations protect intellectual properties—particularly those controlled by Hollywood and U.S. pharmaceutical companies—as a prerequisite to favorable trade agreements.[22] What happened was that the general logic underlying legal decision making had shifted. Previously, if you wanted to extend property protections to things previously unprotected, the burden of proof was on you to justify this extraordinary extension of legal power. By the mid-to-late 1980s, the underlying logic was being reversed. The burden of justification was on the person who wanted to *prevent* legal commodification; the assumption was that, if money could be made, extending intellectual property protection was the logical direction in which to move.

A further indication of the synthesis appeared when, in 1983 MIT social science professor Ithiel de Sola Pool published an influential book titled *Technologies of Freedom: On Free Speech in an Electronic Age.*[23] *Technologies of Freedom* marked the success of the cultural capture of high technology by the economically conservative right wing. If a book with this title had been published in 1935, it would most likely been a celebration of the New Deal and would have focused on things like the Tennessee Valley Authority's rural electrification project. But from 1983 until at least the late 1990s *technology* and *freedom* together would be popularly associated with conservativism. De Sola Pool's book played no small role in making it so. It gave conservatism its needed modernity.

Technologies of Freedom begins with a threatening narrative. The first chapter, titled "A Shadow Darkens," begins by asserting that "for five hundred years a struggle was fought, and in a few countries won, for the right of people to speak and print freely, unlicensed, uncensored, and uncontrolled. . . . As speech increasingly flows over those electronic media, the five-century growth of an unabridged right of citizens to speak without controls may be endangered."[24] De Sola Pool's book made a number of now-familiar moves: it pointed to the coming media convergence due to digitalization as having a potential for enacting a classically liberal utopia of free-speaking abstract individuals, and it demonized the Washington, DC regulatory apparatus as the main threat to the enactment of that utopia. It looked at the past history of free speech through distinctly rose-colored glasses; the five-hundred-year struggle of which he spoke did not include, say, the fact that it was not until the middle decades of the twentieth century that the idea that free speech is fundamental in U.S. law would actually be achieved, in no small part due to efforts of wobblies and other radicals in the 1920s.[25] And it equated free markets, corporate autonomy, and free speech in a common but slippery way, where in one breath the word *freedom* means, say, market competition for local phone service, the next it means abandoning antitrust regulation, and the next it means standing on a soap box in a park proclaiming one's political views.

But the key function of de Sola Pool's book in its day was to describe technological management as a matter of legal rights. Seen as a matter of simple legal process, there's arguably little or no difference between creating a right and engaging in intrusive regulation, and many lawyers would be happy to agree with that assessment. (Until the 1940s or so, people hardly used the term *intellectual property*; what we call intellectual property today was back then thought of by many as a loose bundle of privileges, more in the category of welfare than property.) But putting the word *rights* front and center had a crucial ideological impact at the time. On the one hand, the word *rights* harnessed technology and pro-market enthusiasms to work in broad currents in American culture. Technology and modernity were no longer on the side of planning or the public good or an example of what democratic government could accomplish; they were on the side of rights, and government was their enemy, just as it was the enemy of rights. By 1987, in a book celebrating MIT's media lab, Stewart Brand would cite de Sola Pool's *Technologies of Freedom* as a key inspiration.[26]

The Political Economic Meaning of the Microcomputer

Making sense of a new gadget is not just a matter of figuring out how it works. It has to be given social meaning. Much of this is about standard social variables like status and appropriateness. Carolyn Marvin and others have pointed out that when the telephone was introduced, phone company executives and users alike worried about whether this new technology should be limited to business managers or allowed into the home for other purposes; there was considerable panic, for example, about the possibility that young single women might get their hands on the powerful new device, thereby debasing it.[27] Similarly, when microcomputers first appeared in everyday life, it wasn't quite clear where they belonged in the social order. Was having a computer on one's desk a sign of high or low status? Was this thing with a keyboard for managers or for secretaries? Were they for the home or for the office? Were they for games or serious purposes like budgets? Was using a computer fashionable, a mark of a kind of prowess worth bragging about, or was it simply a routine technical task, like photocopying, something better left to underlings? But as people sought answers to all these questions, patterns were established that suggested something broader: the microcomputer had something to tell us about the relations between politics, economics, and each other. This widely trumpeted new technology became a trope, an ideological condensation symbol, for thinking about big social relationships.

The answers to these questions were not foregone or fully determined by the character of the technology itself. In this period, the French, for example, were also in large and growing numbers typing on keyboards to produce letters on screens powered by microchips—on the Minitel terminals widely supplied by the French phone service. Yet Minitel was a project created largely by the government-owned PTT (Poste, Téléphone, et Télécommunications), and it thus began life as a networked device, for email and retrieving information, rather than as a stand-alone object. Because in France digitalization entered everyday life via a different political economic structure, computing had different implications for meaning and use. The French experienced a connected telecommunications system provided by the government; the box in their homes was a means of access to that larger system and, through it, to fellow citizens. In the United States, in contrast, people experienced an isolated object provided, at least as far as the press was concerned, by capitalist entrepreneurs and inventors. One was a means of access for connecting to others; the other was an isolated box, purely under one's own control. In the broad view, both experiences lent themselves to oversimplifications; the US microcomputer industry sat atop decades of government- funded research in microchip and computer design, and, in France, private industry was deeply involved in manufacturing Minitel terminals and related equipment. Both efforts remained corporate liberal in the sense of business-government interdependency. But such deep-level patterns were not immediately apparent. On the surface, in the case of the microcomputer, subtle and intimate aspects of meaning-creation and the rightward swing in American political economic policy became intertwined.

This intertwining was not purely spontaneous or disorganized. It is true that in the 1970s the idea of mass marketing a general-purpose computer small and cheap enough to be bought by individuals was simply missed by the established corporate players in the industry, and the resulting vacuum set the stage for some small computer firms working off the radar of corporate boardrooms and originating in the hobbyist community, like Apple and Microsoft, to be catapulted into the ranks of major players, with the side effect that the microcomputers arrived imprinted with signs of hobbyist playfulness.[28] But, by the early 1980s, the microcomputer market was no longer being ignored. When Apple and Microsoft were still modest-sized companies faced with a host of competitors like Franklin, Radio Shack, Sinclair, Commodore, Osborne, and Kaypro, American media and some segments of the population were already deeply obsessed by the microcomputer. Even though it lacked the highly coordinated, mission-driven character of other consumer product introductions, the microcomputer "revolution" of the early 1980s was, in a certain way, a highly organized event. What the 1980s lacked in policy manifestos and industry consortia, it made up for with a chorus of mythologizing that resonated from the White House to the media to the local computer store.

Microcomputers as Anticonsumer Products

Scholars in search of the computer zeitgeist of the 1980s have often turned to striking cultural texts like Apple's "1984" TV ad,[29] William Gibson's ur-cyberpunk novel *Neuromancer*, or films with an ominous depiction of computers like *War Games* (1983) or *Terminator* (1984).[30] As fascinating as these dark texts are, it is risky to assume that they actually represented the core of the culture. Apple's "1984" ad was only broadcast once nationally, *Neuromancer* began as a narrow cult classic, and *War Games* and *Terminator* owe more to 2001's HAL than to the rapidly proliferating microcomputers appearing in stores as those films appeared in theaters. If one really wants to get at the significance of computers in the everyday culture of Americans in that period, it might be better to start with something more widespread: the then-novel experience of buying one's first computer. Today, this might seem mundane, but at the time, as millions of Americans went through the experience, it was anything but. Opening the box of a Kaypro, Apple, or Radio Shack microcomputer, assembling the pieces, and having one's first encounter with the then-remarkable experience of typing on a keyboard and seeing corresponding letters appear on a monochrome screen was its own cultural event.

The 1980s brought numerous technological innovations into the home for the first time: videotape recorders, fax machines, and answering machines, to name a few. But these objects were novel only in the sense that more people could afford them; we had all heard of these things before, and they came from companies with which we were familiar. The microcomputer was something in a class by itself. Before the early 1980s, ordinary Americans never imagined they would ever own a computer; computers were thought to be giant expensive things for which an ordinary individual would have no use or desire. So, it represented something quite distinct when swelling numbers of middle-class Americans were suddenly thumbing through computer magazines, deciphering the mysteries of new objects like floppy disks and new concepts like software, and participating in water-cooler conversations about the details of these new machines.

An essential part of the context of the time was the backdrop of a corporate-dominated consumer economy. It has long been observed that in a corporate economy dominated by mass manufacturers consumer products tend to become more and more alike. Jeans, beer, soap, and eventually cars become nearly undifferentiated commodities. The marketing problem thus becomes one of generating an impression of distinction where little exists—hence the habit of slick, jingle- and image-dominated advertising reliant on slogans, celebrity endorsements, and lifestyle imagery, advertising that tells you next to nothing about the product and everything about its cultural associations. The enervating character of this trend over time is one of the weaknesses of consumer capitalism, likely one of the core reasons that groups the world over at odd intervals respond with indifference or hostility to consumer capitalism and its institutions and sometimes go off in search of alternatives.

Many Americans find alternatives to the consumer mainstream by taking up hobbies or avocations that involve constant study of technical details; getting involved in raising purebred dogs, restoring antique cars, or playing the bagpipes requires immersion in a subculture of expert knowledge shared through clubs, personal networks, newsletters, magazines, and books—a subculture in which substance matters, labor is satisfying in its own right, and the activity is beyond the reach of national name brand consumer advertising.

The proliferation of the microcomputer in the early 1980s began as a version of that hobbyist subcultural experience. For the ever growing numbers of individuals thinking about committing the substantial expense necessary to buy a microcomputer—even the cheaper ones cost two or three times the cost of a television set, and some of the popular ones approached the price of a functional used car—there was a world of information available, inviting study, comparisons, and thought, all energized by the sense of excitement of being a part of something new.

The early days of the microcomputer industry thus provided a sharp contrast to the traditional consumer experience and thereby suggested a *capitalist* alternative to conventional consumerism. The technology was indeed new, complicated, and constantly changing, and so the products in question often were indeed distinct. In deep contrast to, say, Coke or Levi's Jeans, concrete information about the products' technical capacities and how they worked was in fact useful and relevant. Buying a microcomputer was not just a matter of picking it up at the store and plugging it in; it involved becoming part of a world of constant reading and discussion, a world in which RAM capacity, microprocessor speed, program compatibility, and peripherals were matters worthy of much attention.

In the 1970s, microcomputer buyers were largely hobbyists and others with technical proficiency connected to technical communities through clubs, school, or work. By the early 1980s, however, much of the purchasing was being done by a wider community of people who had no ongoing connection to such communities—by people with pure curiosity, by small business owners, or by middle level managers who could use the office supply budget and thus operate outside the supervision of central management. These individuals were less able to rely on clubs or other informal networks to get information and thus largely had to rely on the print media to develop an understanding of what these things were and whether or not one should buy one. It is in just such a context that print media can play a powerful role. The early 1980s was marked by an explosion of popular commentary on microcomputers, both in the mainstream press and in a proliferation of new magazines devoted to the topic. Magazines that began in the 1970s as black-and-white narrow-audience newsletters (*Creative Computing, Byte*) adopted glossy formats and swelled with advertising, and a host of newcomers entered the field.

Most of this media coverage took the form of buying advice for novices. As magazine racks began to fill with new computer publications and newspaper articles about microcomputers began popping up with ever more frequency throughout the United States between 1980 and 1983, the tone moved from curiosity to a kind of upbeat urgency, from "what are these?" to "now that you know you're going to buy one, here's how." In March of 1980, for example, the *New York Times* ran a short piece titled "Computers Made to Feel at Home," which was largely comprised of anecdotes supplied by the owners of that new phenomenon, the computer store.[31] The article notes the rapid growth of the fledgling industry— a store owner says, "For the first time since we've been in business, there was a Christmas rush this year." While the article briefly mentions brands and prices— of the Apple II and Radio Shack TRS-80—at this early stage, the article's tone is rather bemused; the article notes that computer stores "have the atmosphere of a friendly club" and points to store employees' enthusiasm for games. To explain what people are doing with home computers, the article relates how (a bit improbably) the children of one new computer-owning family are assigned the task of programming an Apple II to keep household budgets. Another customer is reported to have used his computer "to handicap horses on the theory that it would improve his gambling profits," and another programmed his computer to run his son's electric train.

The following year, the *New York Times* ran a similar piece, "A Bright New World of Home Computers." Here the tone is more confident.[32] Also directed at the first-time potential buyer of a microcomputer, this piece was several times larger and divided into sections: "Doing the Homework," "Tips on Buying," "Where to Shop," and a glossary of computer terms explaining the meaning of words like *printer*. This article expressed little doubt that buying a computer was a serious and worthwhile enterprise. Quoting the advice of university professors, the piece recommends that potential buyers first study computer magazines and books and perhaps take a course. It gets into more detail about technical aspects, explaining, for example, that "programmable memory is measured in the 1,000s of bytes. A byte is the binary code the machine has assigned to every English letter and number. Home computer memories usually range between about 1,000 bytes, or 1K, and about 32,000 bytes, or 32K. K stands for kilo." VCRs were new at the time, too, but this was an entirely different buying experience.

The press was happy to print countless such articles that reported on the constant proliferation of new products associated with the microcomputer. Companies were constantly announcing new models, new companies were constantly jumping into the game, and the third-party industries of software and peripherals were constantly sending out press releases and otherwise hawking their wares. Reporters were happy to write reviews of all these products—this was a rare new journalistic specialty in a tight economy—and companies were happy to buy advertising that would support this reporting. The potential purchaser, after perhaps having his or her curiosity inflamed by someone at the office or an article like the one in the *New York Times*, would then have a plethora of information available at the nearest magazine rack, ready for browsing.

Use Value and Utopia

Significantly, while the press grew ever more confident about the value of purchasing a microcomputer in the early 1980s, what people were actually doing with them remained obscure. In passing, the 1981 *New York Times* piece mentions games, music synthesis, and educational uses, but not in any detail. Between the 1980 piece and the 1981 piece, it is as if the newspaper went from wondering what a microcomputer was for straight to assuming that it had a use—without ever figuring out what that use was. True, the 1980–83 period was the time when various specialized communities were discovering specific uses; academics, journalists, and secretarial staffs were discovering word processing, and small businesses and mid-level managers were discovering the spreadsheet. But it would not be until the second half of the 1980s that the mainstream press would discuss these applications in any detail, and even then it would be more with an "of course" tone, more from the assumption of usefulness than an exploration or explanation of it.[33]

A sizable *Newsweek* overview of personal computers from early 1982 is typical.[34] Titled "To Each His Own Computer," the essay begins:

> Imagine a wordsmith so wise that he can easily comprehend and manipulate more words than most people ever use, then spew them out—with spelling errors corrected—on a high-speed printer. Imagine a teacher with infinite patience who has the devilish ability to spot your weak points and drill you on them, in everything from typing to French. A master sportsman who never tires of playing your favorite game. A researcher who, with a little help from the telephone, can call up the world's great books, the breaking news, the best buy in a local dress shop and the latest breakthrough in kidney research, and bring them all into the privacy of your living room. And imagine all these things wrapped up in a form so protean that even its creators have no idea of its limits—if there are any. There you have it: the personal computer.

This kind of dramatic narrative does not exactly explain what the actual practical uses of the computers available at that time might be; some of the uses described above were more than a decade away, some have yet to be perfected ("a teacher with infinite patience"), and some were grandiose construals of fairly mundane tasks (describing a word processing program as "a wordsmith so wise").

Word processing, the most common use of microcomputers at the time (and remaining so at least until the popularization of the internet), gets two short sentences in this 4400-word article: "Want to write a novel? Load in a word-processing program that lets you and your computer manipulate words on a page, edit the text and print out flawless copy." The article waxes enthusiastic about microcomputers as educational tools—a common selling point in the early 1980s—but says next to nothing about what students are actually doing with the machines. Computers, the article claims, are "emerging as educational tools."

> Classroom 563 is unlike any other at Central High School in St. Paul, Minn. There is no trace of chalk and erasers, school desks or blackboards. Instead, each of the students in Central High's computer lab sits in front of an Apple II plus microcomputer, and the only sounds are the faint clicking of console keys and sporadic beeps from the machines. When the Minneapolis area was hit by a blizzard recently, the students were dismissed at noon and the school cleared in minutes— except for the kids in the computer lab, who refused to go home.

Of the roughly three paragraphs of this thirty-seven-paragraph article devoted to discussion of potential uses, the approach is not a careful discussion that compares traditional against microcomputer-aided methods but an uncritical, breathless rattling off of tantalizing potentials. "The force driving the market," the article opines,

> is the incredible versatility of the machines. A personal computer can be used for an almost limitless variety of tasks. When William D. O'Neill gets home to suburban Washington from his job as Director for Naval Warfare at the Pentagon, for example, he heads for his basement, flicks on his personal computer—and goes to work on his second novel, "The Remorseless Deep," a thriller about submarine warfare. Alan Tobey, owner of the Wine and the People shop in Berkeley, Calif., uses his computer to write recipes to balance the hops, malt and other ingredients for his own homebrewed beer.... Hood Sails in Marblehead, Mass., uses personal computers to design custom sails for yachts in its 25 sail lofts around the world. Or use the computer to generate musical tones and it becomes a musician's instrument. Liberace's composer uses his computer to help with arrangements for the King of Kitsch, and John Cutler, design engineer for the Grateful Dead, uses an Apple II backstage to fine-tune the electronics during the rock group's performances. It can also serve a loftier purpose: the Rolling Stones have an Apple that helps their official biographer store information and write the group's history.[35]

This is not a particularly insightful overview of what was happening at the time or a prediction of what was going to happen. The article does not mention spreadsheets, email, or discussion lists, and while it mentions CompuServe and The Source it certainly contains no mention of the internet. It has no discussion of the obsessive, addicting quality of interaction with computers or of the gamelike quality of much of the computer activity even among those who were not using them to play games.

What the article does is tell the story of the microcomputer, not in terms of its immediate practical uses, but as a narrative of hope, of potential. It offers the reader a sense that he or she has the opportunity to participate in grand, unfinished developments. The vagueness about use was thus rhetorically turned into a strong point; in the absence of hard knowledge about any particular uses that worked, any anecdote could be trotted out as evidence of usefulness.

It was only a month before that *Time's* 3 January 1982 issue declared the computer the "man of the year." The front-cover image was of a grey plaster manikin slumped in front of a stylized microcomputer. The accompanying essay contains more sociological generalizations than *Newsweek's* almost contemporaneous piece, quoting, for example, Toffler's prediction of a postindustrial, edenic, "electronic cottage," while briefly "balancing" such claims with a few quotes from skeptics like Weizenbaum. But the piece is similarly vague about effective uses, choosing to breathlessly list anecdotes that range from the mundane to the improbable, while missing what we in retrospect know to be the key trends at the time; the essay does not mention the spreadsheet, ARPANET, the internet, or Minitel, all of which could have been researched at the time by simply browsing some trade publications.

The Feeling of the Market:
Lone Individuals Exercising Mastery over Objects

There's no reason to ascribe much sociological importance to *Time's* annual publicity stunt, which reflects only the subjective judgment of the magazine's editors, known more for their solid circulation figures than for their prescience or wisdom about current events and social trends. But it's hard not to imagine that a glance at the cover—on newsstands, coffee tables, waiting rooms, and in secondary media commentary as it saturated the culture for a number of weeks—helped solidify, not just the growing sense that computers were somehow deserving of serious attention and expense, but that they were *things*. What was being called to everyone's attention was computers—not computing, not networking, not novel forms of communicating. A computer was, as far as *Time* was concerned, a thing, not a human action. The entire set of activities associated with computing, the still highly concentrated industries of microchip manufacturing upon which the microcomputer industry depended, the pressures to try to turn life into numbers, the legal, political, and cultural battles over control over data, information, and commodification, the shifting modalities of textual production: all this was reduced to a box, a thing, something one saw on a shelf in a store, purchased, and carried away, all for the use of oneself, for an individual, plain and simple. The complexities were all *inside* the box. The microcomputer occasioned the reification of digitalization.

In the 1980s, as a result, these small computers provided an immensely useful object lesson for the proponents of the neoliberal faith in markets. Part of this worked simply because the overlapping communities of politicians and upper level corporate management looked at the rapidly growing microcomputer industry from a distance, watched the rise of new industrial empires like Apple, Compaq, and Microsoft, and saw a powerful argument for reviving the belief that the business world was indeed the product of entrepreneurial initiative. Giant global enterprises from Coca-Cola to GM to General Electric notwithstanding, perhaps the economic world was not dominated by an established interlocking grid of lumbering bureaucratic corporations in league with the government; it was not "the system" that Vietnam era pundits and antiwar protestors complained about. Maybe the business world was simply a world of innovative, risk-taking individuals competing with one another, after all.

The microcomputer thus provided a sophisticated, high-tech glitter to the Reagan era enthusiasm for markets, deregulation, and free enterprise; it became an icon that stood for what's good about the market, giving leaders the world over an extra incentive to pursue neoliberal policies. Gorbachev era Soviet officials have claimed it was the West's astonishing success in market-driven high technology, as much as anything else, that first inspired Soviet leadership to look for new, market-friendly economic models in the 1980s.[36] Neoliberal pundits crowed about the microcomputer industry, and managers and officials who might otherwise be skeptical of the neoliberal theories were given pause.

But it wasn't all at the level of policy and grand theory. For the growing numbers of individuals who took the plunge and bought a microcomputer, there were things about the experience that stood out from the rest of one's life. There was the astonishing effect of an exploding marketplace riding the wave of Moore's Law. In the search for cars or clothing, corporations seemed the same, the prices always rose, and the innovations were largely cosmetic. For microcomputers, however, new prices, new capabilities, and new companies were being announced on an almost weekly basis. The act of buying thus had a sharply different feel.

There also was the contrast with the popular memory of new technologies of the previous two decades, during which new technologies were things like nuclear energy, supersonic airliners, and space travel, each of which had come to involve huge, inaccessible institutions, various degrees of disappointments and dangers, and mysterious complexities that never seemed to clear themselves up, at least in the eyes of the average person. Microcomputers provided a sharp contrast to all those ambiguous or negative connotations. Against that backdrop of technological disappointments, there was something wondrous about having an actual computer on one's desk, in one's home or office, a clean little bit of modern technology that came from small companies with cute names, that produced no roar or smoke and offered no major safety hazard; motorized garden tools were clearly more dangerous.

Finally, for all their little mysteries—in those days, concepts like booting from a floppy disk or typing in cryptic commands were completely opaque to an average user—*those mysteries could be generally conquered.* Most users would eventually learn their way around the basics, and the very process of doing so gave one a growing sense of mastery. Microcomputers were self-contained boxes, it seemed, that with effort could be brought entirely under one's own control.

For any ideology to gain traction in a complex society, for it to become a widely shared form of common sense, it has to be made to *feel* right. This is certainly true at election time, but also more broadly true on the level of the contours of political discourse, on the level that shapes what pundits, journalists, essayists, officials, and politicians perceive as important as they consider legislation, policies, or their next career move. Not everyone who bought a computer came to believe in the free market, of course. Ideological shifts rarely work that mechanically or cleanly. But the experience of reading about, buying, and using microcomputers created a kind of congruence between an everyday life experience and the neoclassical economic vision—the vision of a world of isolated individuals operating apart, without dependence on others, individuals in a condition of self-mastery, rationally calculating prices and technology. A Weberian might call it an elective affinity between the appearance of the microcomputer and the neoliberal faith; a student of cultural studies founder Stuart Hall might call it an articulation.[37] But the point is that, in 1983, even the bearded Marxist professor, using 5¼-inch floppies to boot up his new IBM PC in preparation for working on his latest essay, might at that moment feel a little less in a condition of solidarity with the downtrodden and perhaps a bit more like an ambitious Lockean individual, cutting new ground in isolation. The Marxist's established convictions might prevent him from doing anything with that feeling. But for the much larger number whose political convictions were less fixed that feeling might help them see sense in the neoliberal vision. Maybe markets weren't so bad. Maybe they were even a little bit thrilling.

Beyond Utilitarianism:
The Invention of the Hacker as Romantic Hero

But better that they be a bit thrilling than merely rational. The problem with the kind of autonomous individualism associated with markets, the rational, utility-maximizing individual of utilitarian fable, is that it is so dry; it reduces the free individual to a tedious, calculating shopkeeper. The attraction of the press to Steve Jobs and other microcomputer success stories was not only that they fit the Reagan era entrepreneurial story. If they had been merely successful entrepreneurs in, say, ball-bearing manufacture, economists might have been pleased, but the attention would have been much smaller. Classic utilitarian theory suggests that widespread selfish behavior in a marketplace leads to the betterment of all, but it does not celebrate rebellion for its own sake or care about the colorful details of a capitalist's private life. The story of personal computers being manufactured and programmed by youth in their garages, however, had an added romantic appeal. The microcomputer companies of the early 1980s were young, tweaking the noses of the established computer businesses. The story of young Jobs and Wozniak bonding while building boxes to cheat the phone system in the early 1970s was repeated endlessly in the business press, not because it implied rational, self-interested behavior—it was a college prank, involving petty theft—but because it suggested a rebellious attitude towards the powers that be, expressed through technical aptitude. When Jobs recruited PepsiCo president John Sculley to be CEO of Apple Computer in the 1983, he famously told Sculley, "do you want to sell sugar water for the rest of your life or do you want to come with me and change the world?"[38] In strict marketplace theory, entrepreneurial individuals are not supposed to care about changing the world; the invisible hand is supposed to take care of that. Selling sugar water at a profit is a perfectly rational thing to do. But America's romance with the entrepreneur was a *romance*; flamboyant young characters with dreams of changing the world were much more attractive than mere profit maximizers.

Fred Turner has artfully explained the role of Steven Levy's 1984 book *Hackers* and a subsequent conference in creating—and not just discovering—a particular understanding of hacking as a cultural identity as well as an approach to computer design and programming. Levy, a freelance journalist at the time, had written a book based on interviews with computer programmers who had worked variously at MIT and in the Bay Area and had seized on the colloquialism *hacking* as something significant. Hacking had long been in use to refer to hobbyist-tinkerers who worked with gadgetry for fun, as contrasted with efforts developed according to carefully preconceived plans. Levy traces the term to the

MIT's undergraduate model railroad club in the 1950s. By the 1970s the term had come to be used casually in computer circles to distinguish obsessive and unplanned work styles from those that were rigorous and carefully planned. The general terrain was already laid out; *hacking* referred to the same behavior pattern as Licklider's "self-motivating exhilaration," Weizenbaum's addicted compulsive programmer, and Brand's Spacewars-playing computer hipsters. Following in Brand's footsteps, Levy's depiction of hackers was orthodoxly romantic. They are, he writes, "adventurers, visionaries, risk-takers, artists . . . and the ones who most clearly saw why the computer was a truly revolutionary tool."[39]

Levy added to the discourse both detail and a narrative that gave hackers credit for the ongoing explosion in the microcomputer industry. Subtitled *Heroes of the Computer Revolution*, Levy's book does not mention Engelbart, Licklider, Van Dam, or Xerox PARC; these individuals and institutions were relatively well-funded, embedded in large institutions, and at least on the surface operating according to explicit, rational plans. Instead Levy provides a narrative in which hackers are celebrated as both romantic artists and the true causes of the changes in computing precisely *because* they were romantic artists; they were more easily narratable as such because of their place at the margins of or outside major institutions. Insightfully exploring the cultures of several key moments in the history of computing, Levy weaves a story in which first at MIT, then later in the hobbyist community in Palo Alto, and after that in the nascent early 1980s game-building community, individuals with an obsessive approach to computers advanced the state of the art. Hardly a technical history, instead Levy artfully offers portraits of events and individuals, many of them quite poignant.

Ever since Rousseau, the revelation of internal passions and flaws has worked as a mark of romantic authenticity. Levy's *Hackers* is such a compelling read—it rings true—to a large degree because of his focus on the internal emotional life of hackers. Levy's outline of what he calls the hacker ethic—access to computers should be unlimited, information should be free, mistrust authority, judge people by their hacking, computing can be art—were appealing not so much as political or philosophical statements; as such they would have to be judged as half-baked as best. They were appealing because they were presented as the values of a community struggling with and acting on their internal passions, their shared fascination with tinkering with computers as an end of its own.

To Levy's credit, however, he notices this difference, and at moments shows the tension between the romantic tendency in computing and the rationalizing, corporation-building, profit-oriented imperatives that were asserting themselves in the industry. Levy's heroes are not heroes because they struck it rich but because of their passion and technical contributions. He spends some time in the book exploring the tensions between, say, the profit motive and the ethic of sharing computer code that was quietly building in the programming community. Levy called attention for the first time to young Bill Gate's squabbles with the early computer tinkerers who freely shared his first commercial software efforts and dubs Richard Stallman "the last of the true hackers"; both of these characters would eventually become crucial in the rise of the open source software movement more than a decade later. But, in most of the narratives of the 1980s that celebrated the new computer cultures, the difference between the rational utilitarian form of selfhood congruent with market principles and the romantic form elaborated in Levy was as often as not glossed over or artfully mixed.

Conclusion

In 1983, the same year that de Sola Pool published *Technologies of Freedom*, a curious debate broke out among readers of the countercultural compendium, the *Coevolution Quarterly*, an offspring of Brand's *Whole Earth Catalog*. In the previous decade, the *Quarterly* had in various ways sought to act on the counterculture's egalitarian, utopian impulses. It had offered an issue up to be guest edited by the Black Panthers in 1974, for example. After 1975 all employees received the same pay, and every change in subscription fees was agreed upon by a loosely democratic process of consulting its readers.[40]

Yet, in 1983, some readers complained because the editors had changed the past practice and behaved as most commercial publications do and raised the fee without consulting readers. But this was only part of the changes afoot at the time; in the summer of 1983, Stewart Brand noted all the energy around microcomputers and set out to create a *Whole Earth Software Catalog* and a *Software Review*. Perhaps because writers knowledgeable about computers were constantly tempted by the more lucrative positions with the growing crop of commercial computer magazines, editors of these new projects were offered higher, more competitive salaries. When, in 1984, the *Whole Earth Software Review* and *Coevolution Quarterly* were combined, and the joint publication was named *Whole Earth Review*, the economic egalitarian structure of CQ came to an end. Capitalist practice had been brought back into one of the few relatively visible places in the United States it had been resisted.[41] Simply pointing to the individuals who carried countercultural ideas from the 1960s into the 1980s computer culture cannot explain this shift.

Even though it happened quietly, this was quite a change. One of the most famous moments in the history of the *Whole Earth Catalog* was when, faced with the large profits brought in by the surprising success of the effort, Stewart Brand decided to give the money away to the "community" and held an all-night, come-one-come-all meeting to decide what to do with the money.[42] Brand was never as anticapitalist as some might have imagined him to be; to the extent that it noticed such things, the general ethos of the *Whole Earth Catalog* was that the profit motive was tedious and conformist, not that it was the root of all evil. But by 1984 it was clear that the utopian compass of Brand's publishing realm had shifted, and the inequalities that emerge from market relations were no longer an anathema. By 1990, *CQ* contributors Art Kleiner and Kevin Kelly would go on to help create *Wired* magazine.

This shift could not have been accomplished by, say, simply exposing Brand and his cohort to neoclassical theories of marginal utility. For such a change in the entire terrain of political argument to truly sink in, something has to happen in the gut, on the level of habits of the heart. The microcomputer, newly arrived on the scene, played no small role in that shift. On the one hand, because U.S. computer manufacturers were caught up in visions of automated corporate typing pools and the like, they failed to foresee the demand for general purpose desktop computers and thereby left the field open. This in turn resulted in a very public drama of numerous, small, upstart companies appearing seemingly out of nowhere in a high-tech field and competing fiercely on features and price. While the entrepreneurial narrative of two guys in a garage was an oversimplification, in the case of microcomputers, it was a closer fit than most business behavior in an economy dominated by oligopolistic corporations.

On the other hand, consumers were treated to a particular set of experiences with microcomputers, experiences that themselves stood out from the norm of consumer purchasing. The experience was one of achievable mastery of something one had until recently imagined as impenetrably complex; a generation raised on an image of computers as *2001*'s HAL suddenly found themselves assembling small computers alone in their homes and quickly being drawn into the compulsive character of tinkering with them. Because at the outset these were conceived as largely stand-alone machines, they amplified the sense of computing as something contained within singular object-commodities (rather than as something associated with a system). Microcomputers lent themselves to a vision of oneself and others as abstract individuals competing in a marketplace.

As Levy's *Hackers* knows, the fit between Reagan era entrepreneurialism and the experience of the microcomputer was imperfect in several ways. But ideological shifts are rarely if ever seamless. In retrospect, the confluence of the romanticized stand-alone microcomputer on the popular level with the law and economics and information society movements on the elite level led to a perfect storm of ideological effectiveness, playing a role in launching a period of explosive new growth in capitalist social relations, both across the globe and in the crevices of everyday life and practice in the United States.

4 Networks and the Social Imagination

> One can take issue whether the technical designs that emerge out of that [collaborative] process are the best possible, but it is far more important that they actually come into existence by a process that is open—architecturally open, politically open, that new people come in regularly, that the results are distributed free of charge around the world to everybody. The enormous power of those very simple concepts is very hard to convey to people who have not experienced them.[1]
>
> —Steve Crocker

WHEN CAN A person using a computer be said to be acting alone, and when are they acting with others? At first glance, staring at a monochrome screen and typing arcane commands to connect to and interact over a network is hardly different from configuring a spreadsheet on an Apple II. Both involve esoteric interaction with a machine and lend themselves to an obsessive absorption; both can have the effect of removing one's attention from the physically proximate person in the next room. But, in the United States in the 1980s, for those narrow circles of individuals involved in various stages of the early development of the internet, there were key differences between their experience and the experience of the millions who were encountering microcomputers, and those differences lent themselves to different possibilities for articulation with larger visions. If using a stand-alone microcomputer in the early days lent itself to a feeling of Lockean autonomy from others, using a computer network could have something of the reverse effect; working on a computer terminal connected to a network, particularly over time, foregrounds the social connections embedded in the technology. Anyone who has had to intervene in a discussion list or in a chat room to keep things going smoothly—by, say, giving technical advice to a newbie or by encouraging a flamer to moderate their tone for the sake of group harmony—has had a small taste of this effect. *Attention becomes directed towards the social mechanics of interaction within a system.*

One wouldn't have known it at the time by reading *Time* or *Fortune*, but, looking back, it is now clear that the 1980s was a time of great advancements in computer networking. While the U.S. mainstream was romancing the entrepreneurial tale of the stand-alone microcomputer in the 1980s, outside of the limelight, major developments were taking place with rather different political connotations. By 1980, packet switching was established as a practical means of communication both on the experimental internet and the working X.25 networks that connected banks and research labs. Ethernet (as well as competing token ring and ARCNET) local area networking technology became commercially viable, and the basic underlying protocols for today's internet, TCP/IP, were put into place, tested, and heavily developed. Commercial computer networks like Compuserve and Prodigy were launched, small computer bulletin board systems started to spread, and many university computer scientists began to communicate over the low-cost Usenet system. France led the world into consumer use of computer networks with its nationwide Minitel system, launched by the French post office, that allowed emailing and looking up information on terminals in the homes of citizens. For a substantial segment of the community of computer engineers, networking was near the center of their attention in the 1980s.

Because these events did not enter the broader public eye in the U.S. until a decade later, however, a broad discussion of what happened in the 1980s occurred only after the fact. And, as people have looked back to figure out where this amazing thing called the internet came from, the effort became an opportunity for much hagiography and not a little political mythmaking, both intentional and not. For example, in response to the common (if absurd) mid-1990s habit of attributing the rise of the internet to the free market,[2] Michael and Ronda Hauben published a series of articles and a book that made a strong case that the rise of the internet was due to antimarket, communitarian principles consistent with the 1960s New Left. Howard Rheingold and others in Stewart Brand's circle grafted computer networking onto the New Communalism. And numerous potted histories and timelines of the internet appeared in print and on the internet itself, often reflecting various political inclinations in their selection of details.

Since these early efforts, the discussion has matured, and a more serious historical literature on the evolution of the internet and computer communication has appeared. Works by Janet Abbate, Paul Ceruzzi, and James Gillies and Robert Cailliau have provided much finer detail and careful analysis.[3] But one of the striking things about this newer literature is that, while it is careful not to rush to impose political assumptions onto the historical detail, political questions keep resurfacing. The impact of the internet has been so large, and its origins so distinct, that one cannot help but wonder about the implications of this course of events for understanding politics and social relations.

Here I will focus on a few, illustrative episodes in the evolution of networking in the 1980s, with an eye on what makes them politically unusual and therefore intriguing. Previous work has demonstrated several important points. First, the internet most certainly was not created by two guys in a garage, by small entrepreneurs operating in a classic free market. It was developed inside Hughes's military-industrial-university complex—at a moment when that complex was undergoing significant changes. This fact alone serves an important rejoinder to market fundamentalists and libertarians. But it is also important that the larger development framework was that established by Vannevar Bush. Private corporations were involved in the internet early on and were for the most part always imagined to be central to whatever form the technology would take as it matured; the simple fact that early internet development was funded by tax revenues does not by itself confirm, say, a New Left position with regard to corporations.

Second, part of what distinguishes the early internet development from alternative networking efforts in the 1980s is an unusual culture of informal, open, horizontal cooperation—that very distinct set of practices that are incompletely summarized today under phrases such as "rough consensus and running code," and "end-to-end design."[4] The role of these practices in the history of the internet has become something of a political football; blessed by their genealogical relation to one of the major technological success stories of the twentieth century, they are claimed as supporting evidence by classic corporate liberals, libertarians, democratic socialists, and anarchists alike. It is important to get beyond the simplistic versions of these appropriations. Fred Turner has made the important point that friendly horizontal collaboration among engineers is hardly by itself a guarantor of political democracy broadly construed, and it is in fact historically consistent with autocratic and highly oppressive political structures, like the cold war efforts of the 1950s. And the history is clear that, to the extent there is a politics to internet development, it is not something that can be read off of the political concerns of particular engineers; right- and left-wingers, hawks and doves, all made important contributions, often in cooperation with one another.

This chapter reflects on two instances of a new and unusual set of practices that emerged around 1980, ways of social and technological organization that in retrospect seem relatively politically satisfying and practically effective. First, it looks at the development of new chip design methods in the late 1970s, which led to VLSI (Very Large Scale Integration) microprocessors in the 1980s. While not always listed in the chain of developments that led to the internet, the VLSI chip design process was key to maintaining the momentum of Moore's Law and set the conditions for the parade of ever-improving microprocessors and graphics chips on top of which the internet was built. For my purposes, it nicely illustrates the discovery inside computer engineering of the sheer technical value of attention to social process and to open, networked, horizontal relations. Second, the chapter discusses the much more clearly political economic moment during which the ARPANET efforts were split off from the military and quietly transferred to NSF funding. What is distinct about this remarkable moment is, not just the spirit of openness, but the use of that kind of open collaboration to carefully shepherd a developing network as it passed outside of the cocoon of DARPA funding into a wider, more fraught world of funding by an ever-growing variety of users and sponsors. Theoretically, packet-switched global computer networking could have come to us in any variety of institutional packages, but this 1980s experience of quietly guiding the growing internet into a space between the differently charged force fields of military, corporate, university, and NSF funding left a stamp on the institution of the internet that would have far-reaching consequences.

"We Don't Have to Form Some Institute": The Case of Lynn Conway and VLSI Chip Design

There was never a single justification for seeking to communicate between computers. In the 1960s and 1970s, funding sources were the military and large corporations, so command-and-control uses were favored, such as building a communications network that could survive a nuclear attack, controlling military operations at a distance, or distributed use of centralized supercomputers for scientific research. These are the ideas that dominated grant proposals, committee testimony, political speeches, and mainstream newspaper coverage. But other ideas percolated in the background, such as Licklider's, Engelbart's, Van Dam's, and Nelson's grand dreams of interconnected communication machines.

But the eventual triumph of the ideas of Engelbart was as much a product of surprising experiences with early forms of computer communication as it was a matter of persuasion by a few intellectuals. The most often-mentioned surprise discovery of the ARPANET was the popularity of email and discussion lists; built for command-and-control uses, the ARPANET turned out to be a great way to just chat, and the numbers of emails over the network skyrocketed.[5] These statistics, coupled to the fact that most of the people reading the statistics had personal experience with email themselves, gave substance to the ideas of the likes of Engelbart and Nelson. By the late 1970s, among computing professionals, the idea of using computers for communication between people was no longer abstract; it increasingly had an experiential grounding.

At least as important as the sheer fact of email's popularity was its social tone. Some of this was simply about the informal styles that became customary on email. For example, in 1978, Licklider and a colleague noted:

> One of the advantages of the message system over letter mail was that, in an ARPANET message, one could write tersely and type imperfectly, even to an older person in a superior position and even to a person one did not know very well, and the recipient took no offense. The formality and perfection that most people expect in a typed letter did not become associated with network mes-

sages, probably because the network was so much faster, so much more like the telephone. Indeed, tolerance for informality and imperfect typing was even more evident when two users of the ARPANET linked their consoles together and typed back and forth in an alphanumeric conversation.[6]

It is probably not inherent to computer communication that it encourages informality. It may be simply that, when email started to spread in the late 1970s, the secretaries who were regularly taking hand-scrawled notes on yellow pads and turning them into formal letters were not the ones typing emails. Networked computers were still too rare, expensive, and hard to use to integrate them into the established rituals of the office. The social institutions and expectations that ordinarily lend themselves to formality—secretaries, letterhead, the legal expectations that go with a signed letter—were not operational.

But the informality of online communication was also associated with something more subtle that started to become part of the experience of those using networked computers: the occasional efficiencies gained when working online on technical projects as a group. People often mention the surprising popularity of nontechnical discussion lists in the early days, like Usenet's alt.culture.usenet and alt.journalism.criticism.[7] But the fact is, well into the 1980s, computer communication was predominantly communication about computers; the majority of email and discussion list use was about technical issues.

This might seem like a criticism, but significantly, for the people who designed and built computers, this could be a surprisingly effective way to get things done. An early and influential version of this discovery occurred when Xerox PARC scientist Lynn Conway and Caltech professor Carver Mead collaborated on the development of VLSI design methods for microchips in the 1970s. Carver Mead, credited by Gordon Moore with coining the term *Moore's Law,* was the first to use the methods of physics to predict the theoretical limits of the capacities of microchips. By the early 1970s, these predictions made it clear to Mead and others that individual microchips, especially microprocessors, were destined to become bewilderingly complex. Intel's first microprocessor, the 4004, contained 2300 transistors on a single chip; this was a lot for the time, but it was still something that could be designed by a relatively small team in a matter of months. But, recognizing that this was just the beginning of a trend, Mead foresaw that, as the number of transistors per chip increased logarithmically, this would create new design challenges. How would the complexity of design be handled as the capacity of single chips reached hundreds of thousands, or millions, of transistors?

Lynn Conway, an expert in computer architecture who had made some pioneering innovations at IBM in the 1960s, teamed up with Mead to tackle this problem; as she put it, he was approaching the problem from the level of silicon upwards, and she was approaching it from the level of software downwards. The significant thing about their approach was that they did not set out to design a particular chip or even a particular type of design; they set out *to design a method of design*, a way to make accessible and better organize the process of VLSI microchip design. The problem, as Conway described it in a 1981 presentation, was that

> when new design methods are introduced in any technology, especially in a new
> technology . . . a lot of exploratory usage is necessary to debug and evaluate new
> design methods. The more explorers that are involved in this process, and the bet-
> ter they are able to communicate, the faster the process runs to any given degree of
> completion. . . . How can you cause the cultural integration of the new methods,
> so that the average designer feels comfortable using the methods, considers such
> usage to be part of their normal duties, and works hard to correctly use the meth-
> ods? Such cultural integration requires a major shift in technical viewpoints by
> many, many individual designers. . . . The more designers involved in using the new
> methods, and the better they are able to communicate with each other, the faster
> the process of cultural integration runs. . . . New design methods normally evolve
> via rather ad hoc, undirected processes of cultural diffusion through dispersed,
> loosely connected groups of practitioners, over relatively long periods of time. . . .
> Bits and pieces of design lore, design examples, design artifacts, and news of suc-
> cessful market applications, move through the interactions of individual designers,
> and through the trade and professional journals, conferences, and mass media. . . .
> I believe we can discover powerful alternatives to that long, ad hoc, undirected
> process.[8]

What's distinct here is the extent to which Conway, while working on what she called "designing design methods,"[9] is explicitly talking about *social*, as opposed to purely technical, processes. It's worth emphasizing that Conway is no computer impresario or pundit like George Gilder or John Perry Barlow, who basically make use of the technological for political or social purposes; she is a true engineer working at the cutting edge of her field, giving a talk at Caltech to other engineers. Yet her primary concerns are numbers of individuals, their communication skills, and their culture. She describes her work from this period as a "new collaborative design technology."

Mead and Conway's widely used textbook on VLSI design was not just a summary of what people were already doing; it was carefully thought out to enable more people to participate in the process of microchip design, and was written more with an eye to where microchip design was going than to where it was at the time. Once they had developed some basic ideas about how to simplify the process of design, as Conway put it, "Now, what could we do with this knowledge? Write papers? Just design chips? I was very aware of the difficulty of bringing forth a new *system of knowledge* by just publishing bits and pieces of it in among traditional work. I suggested the idea of writing a book, actually of *evolving a book*, in order to generate and integrate the methods." So the textbook was not just released into the world on its own; it was developed in the context of a series of courses, beginning with one Conway taught at MIT and later extended to several other academic centers of high-tech development, where each course served simultaneously as a way to spread the new ideas and as a way to improve them through tight interaction and quick feedback between everyone involved.

"Perhaps the most important capital resource that we drew upon," Conway states, "was the computer-communications network, including the communications facilities made available by the ARPANET, and the computing facilities connected to the ARPANET at PARC and at various universities." The initial drafts of the textbook used in the first courses, she says, "made use of the Alto personal computers, the network, and the electronic printing systems at PARC" so that they could see the inside of a classroom and be modified based on experience before needing to go through a publisher. Student designs were transmitted over the ARPANET from MIT on the east coast to PARC on the west coast for relaying to a fabrication plant. As the courses expanded to other major universities, the network was used to coordinate the multiple efforts so that all students' projects could be transmitted to PARC for quick and cost-efficient fabrication.

"The networks," Conway observes,

> enable rapid diffusion of knowledge through a large community because of their high branching ratios, short time-constants, and flexibility of social structuring; any participant can broadcast a message to a large number of other people very quickly. . . . If someone running a course, or doing a design, or creating a design environment has a problem, if they find a bug in the text or the design method, they can broadcast a message to the folks who are leading that particular aspect of the adventure and say, "Hey! I've found a problem." The leaders can then go off and think, "Well, my God! How are we going to handle this?" When they've come up with some solution, they can broadcast it through the network to the relevant people. They don't have to run everything through to completion, and then start all over again, in order to handle contingencies. This is a subtle but tremendously important function performed by the network. . . . Such networks enable large, geographically dispersed groups of people to function as a tightly-knit research and development community. . . . The network provides the opportunity for rapid accumulation of sharable knowledge.

Participants in these courses took these experiences and went on to fund start-ups (like Jim Clark, who used his course design to create Silicon Graphics [SGI] and then from there went on to found Netscape) and to build the chips that fueled the continued growth of the computer industry throughout the 1980s into the 1990s.

Other communities of computer professionals were having similar experiences. The pioneers of the Unix operating system, which would eventually come to be the most common operating system on machines that ran the internet, also discovered that there were technical strengths in systems that were designed to lend themselves to communication and collaboration. One of Unix's designers, Dennis Ritchie, famously wrote of the motivation for creating Unix: "What we wanted to preserve was not just a good environment in which to do programming, but a system around which a fellowship could form. We knew from experience that the essence of communal computing, as supplied by remote-access, time-shared machines, is not just to type programs into a terminal instead of a keypunch, but to encourage close communication."[10] Beginning at Bell Labs in the early 1970s, Ken Thompson, Ritchie, and others developed a series of practices that went beyond just particular algorithms, software code, or techniques. Their basic vision of how to approach computer development was distinct. Instead of a company or handful of engineers developing a full-fledged system and then offering it for sale to users—the norm for IBM and other companies at the time—Unix provided what came to be known as a programming environment, where each function was rendered as a separable bit of software that could run on a variety of hardware and was flexible and easily linked to other programs through "pipes." (The famous example here is Unix's search function, grep; instead of building search functions into specific programs like word processors and email applications, grep can search within files from the command line, can be easily connected to other functions, and is thus available for other programs to use; it was an early instance of what became known as a software tool as opposed to a complete program.) Usenet, the legendary early bulletin board system that was the first introduction to computer bulletin board communication for many outside those few that were connected to the ARPANET, was created in 1979 for Unix users at universities to more easily collaborate on Unix-related projects.

Not the least in these efforts was the evolution of the culture of development and governance structures around the ARPANET. The ARPANET was intended from the outset to work across different computer platforms within different institutions, and the procedures for developing the protocols for connecting disparate systems was left largely to the institutions themselves; no single individual or institution was assigned the task of telling everyone else how to interconnect. As a result, a culture and shared awareness developed in the first decades of the internet's life that took into account the need for, and value of, an open, collaborative, nonhierarchical decision-making process. A symptom of this was the creation of the tradition of gently named RFCs (Request for Comments) as the central mechanism for distributing information about the rules and protocols for networking computers over the ARPANET. Out of a few initial meetings attended largely by graduate students, an organization called the Network Working Group was formed (predecessor to today's IETF, which continues to play a key role in the evolution of internet technical standards). This community developed the habit of what has been called "rough consensus and running code," design efforts driven by a loose consensus among expert insiders that is then closely tied to widely shared, practical implementations. Steve Crocker, who as a graduate student wrote the first RFC and who as of this writing remains involved in internet governance, said in 2006,

> One can take issue whether the technical designs that emerge out of that process are the best possible, but it is far more important that they actually come into existence by a process that is open—architecturally open, politically open, that new people come in regularly, that the results are distributed free of charge around the world to everybody. The enormous power of those very simple concepts is very hard to convey to people who have not experienced them.[11]

To illustrate just how deeply (if not widely) the habit of thinking about computer networks as a means to establish horizontal communication for the purposes of technological development had become, it is useful to consider the case of Lynn Conway's move to DARPA in late 1982. Because of the success of her work at Xerox PARC, Conway was recruited to join DARPA to help oversee the newly formed Strategic Computing Initiative (SCI). The project had very conventional Reagan era cold war goals coupled to a Vannevar Bush-style theory of technological innovation; DARPA's official project summary said it would develop technologies, "for such military purposes as aircraft carrier command and control, photo interpretation, and strategic target planning,"[12] while much of the enthusiasm for funding the project in Congress was based on the theory that it was the U.S. answer to Japan's Fifth Generation Computing Initiative, which threatened the United States's technological and, by extension, economic superiority.

By taking this job, Conway was demonstrating that she was no antiwar liberal. (In response to critics, she has said, "if you have to fight, and sometimes you must in order to deal with bad people, history tells us that it really helps to have the best weapons available.")[13] But Conway carried a sense of computers as tools for horizontal communication that she had absorbed at PARC right into DARPA—at one of the hottest moments of the cold war. At the time, she described her goal as fostering collaborative technical development over computer networks, telling a reporter,

> We don't have to form some institute. Wherever people are they can participate. . . . We'll need to have some workshops and some establishing of interfaces among these groups. . . . And then, we'll cook up some network activity. . . . DARPA is to the Department of Defense as the Palo Alto Research Center is

to Xerox. . . . There's a kind of spirit that approaches passion that arises when researchers are forging ahead in new territory. . . . I'm going to try real hard to make some interesting things happen with [DARPA's] money. . . . It greatly oversimplifies to say that we're out to produce a machine. . . . Any one machine is only one point in the design space. . . . You'll see a whole array of technologies and knowledge spin off from the DARPA work. If the work is sufficiently successful, it will have all sorts of applications. . . . [Within 10 years] I imagine that you are going to see a wave of start-up companies as a result of the DARPA-funded research.[14]

The notion that defense research could and should lead to commercial spin-offs was conventional corporate liberalism, in the vein of Vannevar Bush. What is distinct is Conway's style and her enthusiastic description of computer networks as a forum for horizontal collaboration; where her predecessors in the military-industrial complex would have at least gestured to a command-and-control vision of computer development, particularly in a military context, she was speaking frankly and almost exclusively of the value of opening the research process up to relatively informal forms of interaction, to "cooking up some network activity."[15]

The Strategic Computing Initiative is known to some as an expensive failure,[15] and Conway left DARPA after a few years to teach at the University of Michigan. What is significant here is simply that, even in a classic military-industrial context, a computer scientist was speaking a different language about how to make sense of the social relations that undergird technological innovation: "We don't have to form some institute. . . . We'll cook up some network activity. . . [to encourage] a kind of spirit that approaches passion." This is language that would not have been used even by the likes of Engelbart or Licklider in the 1960s.[16] Even deep under the military umbrella, the tone of computer engineering had changed.

In sum, the invisible colleges of computer professionals attached to big institutions like Bell Labs and research universities entered the 1980s already in the habit of thinking seriously about the *specific social organization* of the process of building new technologies, heavily inflected with an interest in creating contexts for effective collaboration and a sense of how hierarchy and institutional allegiances can interfere. Many of the key individuals had specific experiences of cases where open collaboration and the sharing of technical information over computer networks could create efficiencies. Thinking about "designing design methods" was becoming a habit, and easily accessible computer networks were being used as a tool for that purpose. While the rest of the world was dazzled by the stand-alone microcomputer and its association with free market individualism, the communities of computer networkers, who still largely lived out of public view inside the narrow worlds of the university-military-industrial complex, were having experiences that pointed in other directions.

The Internet's Institutional *Annus Mirabilis:*
The Split from the Military in 1983–1984

If Conway exemplified the microstructure of the new network-inflected habits of thinking about the organization of technological development, the fate of the internet in the 1980s reflects a more macro-oriented version of those ways. The practical experiences with the subtle effects of the social conditions of technological innovation were essential to internet decision making in the 1980s. Part of this was simply the discovery of the values of allowing open interconnection. The more participants on a network, the more satisfying the experience of using the network was. Researchers in both public and private settings with access to the ARPANET thus regularly began to allow new participants access to the network, often quietly so as to avoid repercussions from administrators still viewing networking through a command-and-control lens.[17] Alternative, more bottom-up networks like Usenet began to create gateways to the much more privileged and previously exclusive ARPANET.

The story of the internet's relation to the Strategic Computing Initiative illustrates how unusual this kind of reasoning was at the time. SCI was new money, and a lot of it; over its lifetime a billion dollars was spent. The principal creator of SCI was Robert Kahn, then head of DARPA's IPTO and one of the key figures in the creation of the ARPANET in the late 1960s and 1970s. While hopes of science-fiction-like new military applications and besting the Japanese helped wrest this funding from Congress, Conway and the other leaders of the program had a somewhat broader vision in mind with SCI, something that would extend the DARPA tradition of seeding bold, exploratory developments in computing in a way that would advance the entire field. It thus attracted the interest of many researchers throughout the world of computing, from artificial intelligence researchers to solid state physicists interested in new principles for semiconductor design. This was classic Vannevar Bush-styled technological development.

Yet what is striking about the internet in this period was that its leaders *chose* to forgo SCI funding. In particular, Barry Leiner, at the time the program manager of the ARPANET for the Pentagon, specifically declined an invitation to participate in SCI. One might think that internet developers would have jumped at a chance to be involved, particularly given that SCI was under the leadership of one of their own, Robert Kahn. (And anyone who has been involved in the academic pursuit of research funding will note how unusual it is for an ambitious researcher to turn down any funding opportunity.) But Leiner has said that in the early 1980s he was more concerned with avoiding public attention associated with the high-profile SCI than with resources. Funding was less of a problem for the ARPANET at the time, particularly since the desire to interconnect computer networks was now becoming strong among researchers, which meant that many of them—including those involved in SCI—would bring their own resources to the table. SCI's high profile, however, might also bring with it public controversy and meddling, and Leiner's judgment was that such visibility would outweigh whatever benefits could be gained by more funds.[18]

The goal, moreover, was not to be simply secretive or exclusive. In 1980, when ARPANET's Vint Cerf met with a group of computer science professors from across the country, he offered to connect the ARPANET to a proposed academic research network if it adopted TCP/IP protocols.[19] This set the trend towards encouraging open access to the internet, which would become the informal policy throughout the 1980s, leading to dramatic growth fueled, not by lots of government funding, but by an individual institution's desire to interconnect. By the mid-1980s, ARPANET managers Cerf and Kahn were informally encouraging institutions to connect their Local Area Networks—the then-new technology of LANs that interconnected groups of workstations and microcomputers—to the ARPANET. This was a tactic that proprietary systems would be loathe to pursue, but it had the effect of initiating the period of logarithmically increasing internet connection numbers.[20]

The policy towards openness then gradually filtered into the small *p* political world. While avoiding the limelight by staying away from SCI, Leiner also famously shepherded in a new governance structure for the ARPANET that, at least formally, "was open to anyone, anywhere in the world, who had the time, interest, and technical knowledge to participate."[21] Time and technical knowledge were of course major limitations on participation, but those limitations were defined by practical involvement in the technology instead of position in one or another institutional hierarchy. It was a striking bit of openness quietly emerging from near the heart of the military-industrial complex at a very conservative moment in the country's history.

Leiner's decision is a symptom of a subtle sense of the sociology of network innovation and governance that had been evolving inside the community of those working on building the internet, a sense that seems to have influenced much of the decision making in internet development and that contributed to the internet's eventual triumph. Kahn later said that Leiner's "ability to understand how to create social and organizational structures that by their design could motivate individuals to collaborate was at the core of this important contribution" to the creation of the internet.[22] But Leiner was not alone in this ability. From experiences with things like Unix, Usenet, VLSI, and the ARPANET, Leiner and his colleagues had developed an awareness of the value of an informal, committed, open, participatory community environment. That awareness, however, was associated with a corresponding sense of ways that external pressures and agendas could undermine that environment. Corporate profit imperatives, politics, fads, egos, and bureaucratic rivalries could all interfere, at the exact same time that these things generally also provided the context that kept the money flowing. This sense of potential threats and possibilities was driven by accumulated experience and a deep involvement with the technology, not by political inclinations or theories, which meant that it worked at an often not-quite-explicit level. The occasional incorporation of countercultural style, the tolerance of informality, the dodging of hierarchies were not driven by any consensus about 1960s New Left politics; as the case of Conway reveals, the politics of the participants could be quite diverse. But the experiences of the 1960s did become part of the background shared experience of the participants and were drawn upon whenever they might have seemed useful for achieving the goal of widespread computer networking. The community had learned that, in some cases, "we don't have to form some institute."

That shared sense helps explain one of the more remarkable events in the 1980s history of internet development, the separation of the ARPANET into civilian and military parts, which helped lay the foundation for a subsequent civilian, nonprofit, working internet. In October of 1983, the military split the ARPANET into linked but separate military and civilian networks. Most press reports explained this as driven by fears of potential security breaches by hackers breaking into military computers.[23] The nonmilitary, more open network, it was thought, would support militarily significant research; the first reports of the split said the new, nonmilitary network would be named R&Dnet. (It is significant that the name R&Dnet did not stick; the community seems to have understood the "new" network as simply a continuation of the ARPANET and its traditions, while the military branch was seen as something else.) As BBN vice president Robert D. Bressler put it matter-of-factly, "the research people like open access because it promotes the sharing of ideas."[24]

The noteworthy aspect of this move, however, was not that the military requirements of hierarchical command-and-control eventually came into conflict with the growing culture of open collaboration around the ARPANET. That alone would have been unsurprising. Rather, the important point was that the military-centered leaders allowed the creation of a separate, open network for research and development; they kept the internally open system of the ARPA-NET alive rather than simply shutting it down or subjecting it to more severe access limitations. The technology could have been passed on through publications and transfer of personnel; that was the textbook way to conduct a military-to-civilian technology transfer. But in this case the people involved (most centrally Cerf, Leiner's predecessor at DARPA, and his closest colleagues) understood the social commitments and energy that would come from keeping an established, working, and growing network going and managed to quietly carve out a safe space for that network within the pressures that typically come with funding sources. The fact that this event is typically described in the literature and by participants as if it were a technical matter suggests that the sociological sense that was driving decision making in the community at the time was taken for granted. By the end of 1983, a platform for highly effective inter-networking was in place that was protected from the divisive pressures of the profit motive by the mix of military and university contexts and funding, while also fairly well-insulated from the political and technical demands that drove so much of government activity at the time. As if by historical accident, an unusual and what we now know to be enormously productive technological space was created.

The next step in the divorce from the military context was shifting funding to the National Science Foundation, which was accomplished with remarkably little friction or rivalry, no doubt due in part to the astute, under-the-radar style of the key participants. By the second half of the 1980s, a backbone for TCP/IP interconnecting was being constructed called the NSFNET. The internet could from then on be treated as a more generalized research project. To the rest of the world, this all looked like scientists and engineers just doing what they do. That it was something more than that, that in fact an interesting political experiment was underway, would only begin to become apparent at the close of this period, around 1990.

The Information Superhighway: Al Gore, Jr., and the NREN

As 1990 approached, strict market-based economic policy seemed to be on the wane, domestically at least. The stock market crashed in 1987—the first such crash in the United States since 1929—and Silicon Valley was threatened by the Japanese, particularly in the area of memory chip manufacture. The wide-open, unfettered free market was looking a little less inviting and a little more threatening to significant groups of business leadership. As a result, for executives, the business press, and many politicians, a principled hostility to government seemed a little less appealing. Corporations were quietly moving away from the rhetoric of competition and back towards asking for government help to organize and stabilize industries, with calls for regulations that provided "level playing fields" and "regulatory backstops."[25] Some representatives of high-technology industries began calling for government coordinated "technology policy," which was a vague term for the use of government to provide things like tax incentives, research money, and antitrust waivers.[26] Technological progress, many were beginning to believe, could not be left up to the market alone. Explicit forms of corporate liberal cooperation were coming back in fashion.

In the worlds of computing and high technology in the late 1980s, many who were scanning for the next wave—the next best thing after the microcomputer—were finally looking towards networking, but most were imagining things happening in a more collective, centralized way; if there was going to be a digitalized, networked future it was going to be a cooperative project. It was not going to come out of garage start-ups but would involve some form of consortia, private/public coordination and partnerships. Indicative of the trend was the formation of General Magic by a consortium of computer companies in 1987, and the formation in 1989 of the Computer Systems Policy Project, a lobbying group made up of the CEOs of ten computer manufacturers, including AT&T, Digital, Hewlett Packard, and IBM.[27]

Similar impulses were driving efforts in networking. As the network—now increasingly called the internet—continued to grow and possible commercial uses began to come in sight in the late 1980s, things seemed to be going according to formula. In 1988, computer scientist Leonard Kleinrock chaired a group that produced a report, "Toward a National Research Network"; this report caught the attention of, among others, Al Gore, Jr. In May 1989, the Federal Research Internet Coordinating Committee, released a "Program Plan for the National Research and Education Network," which proposed, after an initial government investment in a high-speed network backbone to major computing sites, that subsequent stages "will be implemented and operated so that they can become commercialized; industry will then be able to supplant the government in supplying these network services."[28] That same year, physicist and former IBM vice president Lewis Branscomb teamed up with Harvard-trained lawyer Brian Kahin to found the Information Infrastructure Project (IIP) at Harvard's John F. Kennedy School of Government, with funding from a rich mix of foundations, government agencies, and corporations.[29]

The tone was high minded and acronym-laden—National Research and Education Network quickly became NREN—with an emphasis, as per the title, on applications like education and scientific research. "The NREN," it was proposed,

> should be the prototype of a new national information infrastructure which could be available to every home, office and factory. Wherever information is used, from manufacturing to high-definition home video entertainment, and most particularly in education, the country will benefit from deployment of this technology.... The corresponding ease of inter-computer communication will then provide the benefits associated with the NREN to the entire nation, improving the productivity of all information-handling activities. To achieve this end, the deployment of the Stage 3 NREN will include a specific, structured process resulting in transition of the network from a government operation a commercial service.[30]

Drafts of legislation began circulating in Congress, proposing federal funding for a network that would "link government, industry, and the education community" and that would "be phased out when commercial networks can meet the networking needs of American researchers."[31]

Many readers will remember all the talk about the "information superhighway" in the early 1990s. Because of the rich mix of political and economic energy to which the phrase became attached, it developed a lot of momentum. Politicians sought to ride on its coattails, and industry factions began to try to capture it; phone companies claimed they could provide the information superhighway, provided the government stayed out of it, thank you, and the cable industry countered by politically correcting the name of their newest technology from "500 channel TV" into "cable's information superhighway."[32] *Information superhighway* became so common it sprouted its own metaphorical universe, involving phrases like "road kill on the information superhighway."[33] It's easy to forget, however, that for the first few years of this buzzword's flourishing, the information superhighway was not necessarily the internet.

The phrase *information superhighway* has been around since at least the early 1980s and the metaphor of an information highway for at least a decade before that.[34] But around 1990, *information superhighway* began to take on a very specific life inside the political circles of Washington, DC. At the time, the U.S. economy was floundering, and the administration of George H. W. Bush was looking increasingly helpless on the economic front. *Fortune* magazine sniped that "the President has been disengaged, reactive, and inarticulate" on the economy.[35] The Democrats in Washington sensed an opportunity; the slogan "it's the economy, stupid" would soon prove devastating to Bush in the next election. But the problem for the mainstream Democrats was finding a way to differentiate themselves from the Republicans without opening themselves up to the label of tax-and-spend liberals that had been used so successfully against them in the previous decade by Ronald Reagan.

In the 1950s, Senator Albert Gore, Sr. had made a name for himself by shepherding in the interstate highway system, which gave a huge boost to the auto industry and the economy in general while profoundly shaping American life and culture around the automobile. It was one of the most successful and beloved massive U.S. government building projects of all time, a triumph of corporate liberal habits. To this day it stands largely above criticism. No doubt this rousing success was somewhere in the back of then-Senator Albert Gore, Jr.'s mind when, starting in 1988, he decided to get involved in building computer networking in the name of research. Gore, Jr.'s inspiration was to link up with various proponents of advanced computer networking in the engineering community, sponsor legislation that funded the development of a state-of-the-art computer network of networks, and call the project the "information superhighway."

The idea pressed several buttons at once; the high-tech industries, battered by Japanese competition and nervously groping for the next wave, looked favorably upon this modest kind of government investment, which after all could save them the cost of a lot of high-risk R&D and perhaps shield them from overly intense competition. Because the project was wrapped in the glamorous aura of high technology and a positive vision of the future, Democratic politicians, like Gore, Jr. himself, could use this safely as a model of "good" government intervention, undermining the Republicans' efforts to maintain power by associating Democrats with government bureaucracy and excess. And it appealed to a kind of economic nationalism; by 1991, a Congressman argued for government involvement in the creation of a U.S. broadband network by saying "the Japanese will have an information superhighway by the year 2005 and the USA won't."[36] Small wonder, then, that Gore, Jr.'s bill moved calmly through both houses of Congress and was signed by President Bush in 1991, providing for 2.9 billion dollars over five years for building the NSFNET.[37] At the time, Al Gore noted, "in many ways, this bill is very unusual. I have been working on this bill for more than 2 years, and almost no one has said a discouraging word about it. Instead, I hear enthusiastic support in many, many different quarters—within the administration, in the telecommunications industry, in universities, in the computer industry—among researchers, teachers, librarians, and many others."[38] And then in 1992 the election of the first Democratic president in more than a decade seemed to make the political climate favorable for this kind of public-private effort. This looked like a classic implementation of Vannevar Bush's corporate liberal principles for technology development.

The Public/Private Problem and Com-Priv

But the Bush philosophy does not always lead to the linear, orderly process it is sometimes imagined to.[39] Corporate liberalism mixes private and public, and for all its historical effectiveness, that mixing creates a substantial grey zone where the rules are unclear and asks the polity to take a lot on faith about both the good motives and the wisdom of the individuals at the center of this movement between the two worlds. And it inevitably raises the question, why should private companies and individuals profit from publicly funded research? Why is this not government favoritism?

Bush himself squabbled with President Truman and Congress in 1945 over the exact form that the National Science Foundation was to take. One Congressional bill, for example, proposed that all patents for the government-funded research be retained by the government, whereas Bush favored protecting private patents out of concern for maintaining flexibility and autonomy.[40] Bush's approach was based on a deep faith in the capacity of scientists, engineers, and other experts to overlook their own selfish interests in the name of reason and progress. The Congressional proposal, by contrast, was based on a more transparent, skeptical logic. The fact remains that the process of transfer and public/private cooperation in general involves neither a Lockean market nor a public process dedicated solely to the public good. It is a movement between different worlds that operate by different rules. There is no getting around the fact that research efforts paid for at least in part by public tax money come to serve the interests of those who are making a private profit.

These tensions were laid bare on one of the more lively and revealing public political economic discussions of the early 1990s, a now-legendary discussion list called the Commercialization and Privatization of the Internet—com-priv for short. The community of network experts, having spent the 1970s and 1980s simultaneously developing the technology, discovering its pleasures, and learning the value of an open approach to its coordination, did what to them was the obvious thing; when faced with the sociopolitical complexities of making the internet into something broadly available, they established an electronic discussion list, open to all with the means and interest to sign up—which at the time, was still a relatively narrow circle.

Com-priv was initiated by Martin Schoffstall, a long-time participant in the Internet Engineering Task Force who had recently founded a company called PSI to offer access to the internet on a commercial basis. Opening his initial post to the list with the address "GentlePeople," Schoffstall laid out some questions: is the open, casual, RFC-based process of decision making adequate for a commercial environment? What will be the relationship between existing, tax-funded, nonprofit network providers and commercial newcomers (at that time, PSI and a company called Alternet)? What happens when commercial activities start taking place on noncommercially funded systems? Schoffstall concludes the post in a way appropriate to the inviting, informal tone that had become the norm in behind-the-scenes internet decision making: "Come let us reason together. . . . Marty."[41]

Some of the discussions that followed remained technical (for example, "How long does the UNIX password encryptor take on an 8088? Is it faster or slower than a PDP-11?").[42] But one of the striking things about the list is how much of it is devoted to working through policy issues; engineers found themselves thoughtfully debating fundamental principles of political economy. Much of the initial discussion began around something called the acceptable use policy (AUP).[43] After the transition from a defense department umbrella to the NSF, the network had evolved around the central, NSF-sponsored TCP/IP backbone called NSFNET, which was then connected to a variety of regional networks, most of which were nonprofits and often leased equipment from for-profit companies. High-tech corporations like BBN and Hewlett Packard, with their interests in networking and computer research in general, had various kinds of connections. PSI and a company called Alternet had begun offering access to the system on a commercial basis. The NSF portion of the network, however, was governed by a policy that said the network should be used only for appropriate research and education purposes.

With regards to the Acceptable Use Policy, Schoffstal asked, "How does one constrain use of federally subsidized networks . . . from doing commercial things?" Allen Leinwand, then a network engineer working for Hewlett Packard, elaborated on the problem:

> This question has plagued us here at HP for some time. . . . Suppose that HP connects to AlterNet (we have not . . . yet) and we now have the ability to pass commercial data across AlterNet legally to company X and company Y who are HP business partners. We are already considering the idea of subsidizing our critical business partners with the funds to connect to AlterNet when we do. . . . The main problem is how do you convey to about 90,000 employees that it is legal to conduct commercial business with IP based services to company X and Y because they are on AlterNet, but don't do it to company Z because they are only on BARRNet (the Bay Area public regional)? . . . I cannot really envision a network tool which intelligently decides what data is for commercial use and what is not. How do we distinguish between HP divisions working with the OSF across NSFNET (which IS legal) and the same division (or machine!) sending data to company Z?[44]

The subsequent discussion of this issue came up with more examples and explored different possible solutions. A purely technical solution was discussed, where different uses get coded into the network routing system, but it was generally deemed impractical because of the already quite blurry lines between nonprofit and for-profit activities on the network. Something that involved collective human decision making was needed. The problem was, in essence, political.

Political, but not polemical. The discussion on com-priv made the goal of a fluid, easy-to-use, open, and reliable network a priority above all else. Schoffstal, who had recently stepped into the role of an internet capitalist, wrote,

> What PSINet has been doing (and from all appearances what ALTERNET has been doing) is working with industry and not upsetting the stability of the non-profit mid-levels from providing service to the non-profits and academics. That non-profit infrastructure seemed pointless to hurt since too much of the US is incredibly dependant on it. . . . Now when the non-profits provide service to industry is where we get into a sticky philosophical/legal/taxation areas.[45]

Neither Schoffstal nor others tried to resolve issues by adopting principled anti-business or antigovernment ideological positions that are so common in other public debates. Parts of the system that worked, in this case run on a nonprofit basis, were best not to be interfered with, even by for-profit entities. The approach was highly pragmatic.

But there was the matter of what Schoffstal called "philosophical/legal/taxation areas." Pragmatism around strictly technical matters is one thing, but pragmatism when it comes to the murky world of political and institutional structure is quite another. In the latter there is no getting around the fact of political and social choice; decisions will have to be made that will allocate and shape the distribution of power and resources, and no legal or technical necessity will dictate the form of those decisions in a completely neutral way.

Conventional corporate liberal decision making in the United States has generally dealt with these moments by couching things in the language of expertise, bound together by reference to the national interest or public good. When Herbert Hoover set out to organize the new technology of radio on a corporate for-profit basis in the 1920s, he gathered together a mixture of captains of industry and engineers and used the language of "the public interest, convenience, and necessity" to justify the creation of an administrative agency (predecessor to today's FCC) that proceeded to use government legal power to allocate radio frequencies in a way that favored large, well-funded, commercial operations. When tax money was used to create the interstate highway system in the 1950s, the legitimating language was that of national defense. In each case, the public language was highly formal and bureaucratic.

The tone on com-priv was something different. Instead of falling back on the authority of expertise and institutional hierarchy, there was an explicit small *d* democratic impulse: "come, let us reason together." That impulse was leavened by small gestures developed in past pragmatic experiences with such forms of decision making. Most of these gestures were tokens of informality: first name modes of address, occasional colloquialisms and personal details, and the use of self-mockery. Schoffstal, in describing an individual whose "position was a bit stronger than I would have taken" quickly adds a parenthetical aside, "(hard to believe for some of you)." All these gestures worked to soften personal sharp edges, generate a tone of informal solidarity, and facilitate group process.

United by the common goal of a functioning network, then, the community on com-priv was using what had worked for them in technical areas—free flowing, horizontal, electronic communication—to self-consciously deal with issues that were both philosophical and political. Tinged by (if not fully committed to) a post-1960s suspicion of established, formal institutional habits and by a corollary trust in informal directness, they set out to negotiate the blurry terrain between government and for-profit rules of operation in a manner that at that point in history was unique. They took what worked in a technical context—rough consensus and running code—and set out to apply it to matters that were becoming increasingly political.

In the broader political world, however, other habits dominated. In December of 1992, President-elect Bill Clinton convened a Conference on the State of the Economy, having made fixing the economy a centerpiece of his campaign. The conference brought together a blue-ribbon group of experts and corporate chieftains. At this point, the rhetoric of the information superhighway was in full swing, and so it was on the agenda, which gave Vice President-elect Al Gore, with his experience in setting the stage for NREN, a chance to shine on one of his favorite topics.

The *New York Times* quoted this exchange between Gore and AT&T chair Robert E. Allen. Allen said,

> A focus on infrastructure, including information networks, commercial networks which are interconnected, interoperable, national and global, needs to be encouraged. I have some points to make about who should do what in that respect. I think the government should not build and/or operate such networks. I believe that the private sector can be and will be incented [sic] to build these networks, to enhance them and make it possible for people to connect with people and people with information any place in the world.
>
> I do think, however, that the government role can be strong in the sense of first, increasing investment in civilian research and precompetitive technologies. Secondly, supporting the effective transfer of that technology to the private sector. Thirdly establishing and promulgating technical standards, which are so important to be sure that networks and devices play together, work together, so that we have the most efficient system in the world. And incentives for investment and research development, job training.

Gore replied:

> I fully agree when it comes to conventional networks and the new networks that your industry is now in the process of building. But with an advanced network like the National Research and Education Network, it does seem to me that government ought to play a role in putting in place that backbone. Just as no private investor was willing to build the interstate highway system, but once it was built, then a lot of other roads connected to it, this new very broad band high capacity network most people think ought to be built by the federal government and then transitioned into private industry. You didn't mean to disagree with that view when you said government should play a role did you?

To which Allen responded, "Yes I may disagree."[46] The next day, *USA Today* reported on the exchange under the headline "AT&T's Allen Feuds with Gore."[47]

There's a point of view that sees this dispute as merely technical. Gore and Allen both agreed with the basic Bushian approach in which government funds initial research and then hands things off to private industry for practical development; the question was simply about whether the initial backbone for a proposed high bandwidth network should be government-created and then handed off to industry or built by the private sector from the beginning. Brian Kahin, who was already staking out a coordinating role in this effort through the Kennedy School's Information Infrastructure Project, later complained that the Gore-Allen exchange "confused the issue." As far as he was concerned, everyone already assumed that the internet would be "opened up to the private sector." The network already was being built from the bottom-up by private entities—companies and universities—providing local area networks and workstations; the government was just providing some help near the center, with some "top-down subsidies." Private companies like IBM and MCI had already gotten government contracts to build significant parts of the technology, and in some cases, according to Kahin, they gave bids below costs, presumably because they viewed this as an R&D investment.[48] For someone like Kahin, all this was simply reasonable coordination of public and private efforts; concerns about the public internet being "turned over" to private enterprise were much ado about nothing.

But ado there would be. Ambivalence about the appropriate relations between government and for-profit enterprises are woven into the American soul. The Gore-Allen exchange would be just the first in a sporadic series of public squabbles revealing uncertainty over the boundaries between the private and public status of the internet in the years to come. Some of these squabbles would be mounted by political activists of various types; anticorporate activists would complain about the theft of public goods, and economic conservatives would try to prove that Al Gore (and public funding) had nothing to do with the success of the internet. But, even for those with less specific political agendas, the Gore-Allen exchange raises deep questions: who decides these things? Is it right that someone like Allen, with obvious corporate loyalties, be allowed such influence over this level of decision making? And was Al Gore's vision of a supportive government investment as cleanly rational as he made it out to be?

Looking back on his leadership in developing legislative support for the NSF-NET, Gore said during the 2000 campaign for U.S. president, "I took the initiative on the internet." This statement was then attacked in print by libertarian *Wired* magazine reporter Declan McCullough and eventually twisted by various Republicans into the sound bite that Gore said he invented the internet.[49] From there it went on to become a favorite joke of late night comedians and a punch line in a TV pizza ad. It was a false slur, and it was irresponsible of reporters and politicians to repeat that sound bite up to the end of the campaign; it seems plausible that the "Gore said he invented the internet" quip did at least as much damage to Gore's final vote count as Ralph Nader.

But what's important about this episode is that, while the sound bite was factually untrue, it was funny. And it was funny because it appeals to a common skepticism about the orderly, managerial mode of thought associated with technology policy like Gore's. As far as Washington was concerned, the NREN was consistent with traditional corporate liberal policy; it was to be a technology test bed, something that would provide innovations that would eventually be implemented and broadly deployed by the private sector. And it would develop on a national basis, neatly coordinated by orderly consortia of established corporations like IBM and AT&T, perhaps eventually linking up with equally orderly systems developing in other nations around the world. It was all very highminded. The information superhighway predicted by Gore's NSFNET initiative would be used by scientists for sophisticated research and perhaps as a kind of electronic library where thoughtful patrons would quietly and studiously gather useful information.

Gore did take the initiative on the internet, but what he had in mind was hardly the chaotic, explosive phenomenon that would soon be conveying a cornucopia of pornography, pop culture, conspiracy theories, and irrational exuberance throughout the globe. And he did not have in mind the kind of gently self-mocking, open, deliberative process that was taking place on com-priv.

Conclusion

To what extent did the habit of attention to social process that evolved in the 1980s matter in the history of the internet? In the early 1980s, the eventual triumph of the internet, especially of the TCP/IP protocol developed for the ARPANET, was by no means foreordained. Numerous other experimental networking efforts were running at the time, such as Britain's SERCNET (Science and Engineering Research Council Network), the French Cyclades network that inspired some of the techniques that had been brought into TCP/IP, and proprietary packet-switched internetworking systems provided by computer manufacturers such as DEC's DECNET. Most prominently, the X.25 networking standard, which was working commercially at the time and had the support of international bodies and telecommunications carriers, was considered by many to be the obvious wave of the future. In France, the PTT brought networking to common people with Minitel, whereas in the United States, networking remained largely buried away inside universities and the military-industrial complex for most of the 1980s. Minitel and other teletex systems seemed to many like the obvious way to bring digital communication to consumers. The conventional international body for setting technical standards, the ISO, was working hard at developing a global packet switching protocol called OSI, in theory more advanced than the internet's TCP/IP.

Moreover, by 1980, the cold war consensus that had given urgency to ARPA/DARPA in the 1950s and 1960s was gone. When the newly elected Reagan administration engaged in some high-tech saber rattling in the form of the Strategic Defense Initiative, the computer networks at Xerox PARC hummed with expressions of opposition, and a network discussion list was formed in October of 1981, which in 1982 lead to the creation of Computer Professionals for Social Responsibility (CPSR). The group became internationally prominent for their opposition to Reagan's SDI (popularly known as Star Wars), which was based on the theory that high technology—backed by the latest computers—could be used to shoot down enemy ICBMs. The program was not only destabilizing and likely to be perceived as a belligerent threat, CPSR argued, but it would not work because of the fallible nature of computing. In sum, the full intensity of the day's most heated political disputes had appeared within the still tiny world of computer networking. Some, like Conway and Kahn, still felt it was legitimate to engage in weapons-related research, but that choice was no longer taken for granted. The easy combination of social with military visions characteristic of Licklider's early work could no longer be assumed.

So why did the internet succeed? As with all technological standards, one has to allow for a certain amount of dumb luck: of accidents of timing, economics, or politics. The Betamax vs. VHS story is a favorite example; the adoption in the United States of the mediocre NTSC standard for color television in the early 1950s is another. One needs to take seriously the possibility that many or most of the reasons for the internet's eventual success over other standards and systems are simply ones of historical accident. One commonly offered explanation for the success of the internet's TCP/IP standards, for example, is simply money and timing. Buoyed by defense department funding into the 1980s, the internet's TCP/IP standards reached a critical mass of effectiveness and users at a time when other standards, notably OSI, were still struggling to achieve financial support and political and technical stability. Had some historical accident slowed the internet down or sped OSI up, the argument goes, internetworking may have followed a quite different path.

That said, in retrospect, clearly something was done right in the 1980s, and that something was not purely technical; it was not just the adoption of the specific technologies of packet switching and end-to-end design in the abstract. There are some political lessons to be learned from the success of the internet.

Most obviously, the internet was not turned into an effective communications medium by two guys in a garage, by corporate leadership, or by real or imagined market demand.⁵⁰ The free market visions and entrepreneurial fables that caused American culture to obsess over the microcomputer in the 1980s also created a blind spot that for the most part rendered invisible the collaborative, social institutions that created the internet. To a large degree, the context for the creation of the internet was defined by nonprofit institutions using government funding, with participation by the private sector, by individuals consciously working in a collaborative mode who valued freely sharing information and code within the technical community. Whatever one makes of those processes, they were definitely not entrepreneurial or market driven in any obvious senses of those terms—and the broad public ignorance of that fact would have substantial consequences in the 1990s.

But what, exactly, were those processes? One can find individuals of many stripes among the computer pioneers of the 1970s and 1980s who still draw divergent political conclusions from the period. Hauben and Hauben's *Netizens*, written during the 1990s as a critique of the marketplace enthusiasms of the time, would have the lessons of Unix and ARPANET development point towards the value of an antimarket communitarianism. Yet Steve Crocker, for example, has been involved in several commercial start-ups; he apparently does not see a fundamental incompatibility between valuing a system in which "results are distributed free of charge around the world to everybody" and private enterprise. If there is a dominant view amongst engineers, it is probably still some version of Vannevar Bush's: technological innovation requires a mix of public and private efforts, where the pattern is for nonprofit institutions to do basic research and perhaps establish shared frameworks and protocols, which are then passed on to private industry for development into working systems. Conway, for example, has repeatedly emphasized the ways that her work at PARC and later have fed into the creation of private start-ups and qualifies the collaborative sense of what was going on by calling it a "new kind of collaborative/competitive environment."⁵¹ And, in the end, it must be acknowledged that there is a relative autonomy of engineering; engineers who work together are focused above all on getting complicated things to work, and that imperative seems to be able to create cooperation among engineers with quite different social and political proclivities.

Part of what happened was that, for key networking pioneers in the 1980s, attention to social, institutional, and political relations became increasingly folded into their technical concerns. In roughly the same way that learning a foreign language often makes one more aware of the grammar of one's native language, the act of building computer networks made one more aware of social relations. Diverse individuals and institutions established effective technical gateways between networks, developing protocols that allow different machines to interoperate, and similar tasks brought many of the material complexities of communication to their awareness. This was not without its pleasures, furthermore. Over time, the network grew, creating a situation where logging in always brought with it the possibility of new members of the discussion list, new connections and gateways to other networks—to new people, new social contacts. As the networking pioneers became more experienced with these fundamentally social aspects of networking, some of them eventually became involved in national legislative affairs, international regulatory debates, and some unique struggles inside the military-industrial complex. Out of the broad public eye, they laid the foundations for what would become today's internet, both on the technical level and on the level of political habits of governance.

In 1983, the to-that-point largely informal, open, democratic culture of early internet governance and technological development was given crucial institutional support. What could have been a minor aberration in the history of military-funded computing was instead nourished and encouraged so that it would eventually provide the technology and ethos that would lay the foundations for today's internet. It would be another decade before the internet would explode on the global stage and overtake numerous corporate efforts to popularize computer communication, becoming the vehicle for the realization of the long dreamed-of multipurpose global network. But the Department of Defense's 1983 decision to support a packet-switched network free from military command hierarchies was a key moment in creating the conditions in which open TCP/IP packet-switched networking was able to gradually evolve into the all-pervasive internet.[52] A community of researchers already attuned to the values and pleasures of decentralized computer networking and informal, horizontal means of governing the evolution of the technology through RFCs and voluntary committees was given the funding and legitimacy necessary to further cultivate the network outside the immediate pressures of both military hierarchies and the profit imperative.

The previous discussion suggests that the decision was as much a product of the culture of the community of computer engineers as of specific institutional needs or designs. A large part of what would come to triumph with the explosion of the internet is a particular vision of what computers are—writing technologies, devices for manipulating and communicating symbols among equals—associated with a particular kind of informal social vision, a set of beliefs about human social relations. That vision first appeared in the late 1960s at least partly under the influence of the 1960s counterculture, quietly grew through the 1970s, and by the early 1980s had sufficient influence to shape decision making within the Department of Defense. That vision is neither monolithic nor self-evidently politically generalizable. But the 1983 decision in particular illustrates the extent to which key decision making about network organization has been shaped as much by the cultures of the relevant communities of innovators as by macro-structural and economic forces.

5   The Moment of *Wired*

When it comes to smashing a paradigm, pleasure is not the most important thing. It is the only thing. [The web browser] Mosaic is not the most direct way to find online information. Nor is it the most powerful. It is merely the most pleasurable way, and in the 18 months since it was released, Mosaic has incited a rush of excitement and commercial energy unprecedented in the history of the Net.

—*Wired*, Oct. 1994 (10 months before the Netscape IPO)

Revelations in the Cubicle:
White-collar Computing in the Early 1990s

Recall—or, if you are young enough, imagine—what it was like to go online in the early 1990s. At the time, desktop computers had recently lost their novelty and become a routine part of office life. Word processing had, in the preceding five years, become a standard secretarial skill, and a new desktop computer was a standard part of an academic job offer. The desktop computer had become just another part of office routine, like the photocopier.

In most offices, however, people who used email were still a small minority, and web browsing was unknown. Those who had experimented with email a bit had done so typically within specific, confined worlds like CompuServe, Prodigy, local bulletin boards, or one of several restricted academic or corporate networks. Going online at the time was thus technically possible with the computers that were on the desks of journalists, academics, and other professionals, but it was a little out of the ordinary. If you weren't a computer professional, it was something you did out of curiosity; it took a substantial amount of time and was unlikely to yield much in the way of immediate practical value. For the vast majority, communications that mattered still happened exclusively on paper or on the phone. If you went online you knew that most people around you did not.

Going online typically required purchasing and plugging in a roughly paperback-sized modem (computers did not routinely come equipped with them). The modem had a bank of mysterious flashing red lights, and using it involved installing, configuring, and then running a terminal program, typing commands, listening to the squealing modem, and typing in another cryptic series of commands and passwords. There was no pointing and clicking yet in the online world. Just getting it all going was at least a forty-five-minute time investment. And then figuring out what to do once signed on was a further challenge. Gateways between computer networks were still being constructed. As a result, to send, say, an email from the BITNET network—then common at less technical universities—across the still limited internet, the email addresses had to be sandwiched between quote marks and prefaced by IN%—thusly: IN%"T_STREETER@ uvmvax.uvm.edu"—and this technical detail was not easy to find out.

But, once you mastered such arcana, you could then enter into a secret world.

This was the context in which a message appeared on a number of discussion lists in February 1993, prefaced with the following:

> From:IN%"TNC@GITVM1.BITNET""'TECHNOCULTURE' discussion list" 22-FEB-1993 11:48:56.39
>
> To:IN%"T_STREETER@uvmvax.uvm.edu""Thomas Streeter"
>
> Subject: John Perry Barlow meets the spooks
>
> Folks,
>
> This lovely missive came from SURFPUNKs (subscription info below). The idea of JPB giving an invited address on technology to the intelligence (sic) community is just soooo sweet. And it's a good speech, too.
>
> Larry Hunter

The bulk of the message was the text of an address given a few months before, in December 1992, by Electronic Frontier Foundation (EFF) cofounder John Perry Barlow to a conference on National Security outside of Washington, DC.[1] As the message made clear, many members of the U.S. intelligence community (that is, the CIA, NSA, FBI) were present. Barlow's agenda as EFF representative was to educate this community about the value of protecting free speech and privacy in the digital realm.

Ordinarily, when speaking to a skeptical audience, most of us are likely to adopt a careful, formal, conformist rhetorical strategy. We would downplay our disagreements and differences and represent ourselves as having deep respect for the audience members. Barlow, however, began his talk this way:

> I can't tell you the sense of strangeness that comes over someone who earns his living writing Grateful Dead songs, addressing people who earn their livings as many of you do, especially after hearing the last speaker. If you don't appreciate the irony of our appearing in succession, you have no sense of irony at all. . . .
>
> The reason I am here has absolutely nothing to do with the Grateful Dead. I'm here because I met a fellow named Mitch Kapor in 1989. Despite obvious differences, I felt as if we'd both been up in the same saucer or something . . . that we shared a sense of computers being more than just better adding machines or a better typewriters. We saw that computers, connected together, had the capac-

ity to create an environment which human beings could and did inhabit. . . . The people who share this awareness are natives of the future. People who have a hard time with it may always be immigrants.

When Mitch and I saw that computers had created a place, we started asking some questions about what kind of place it was. . . . We decided to name it Cyberspace, after Bill Gibson's description of a futuristic place rather like it which we found in his novel Neuromancer.[2]

Here is a central example of some habits of talk and thought that would soon be moving into the mainstream with enormous impact. Barlow was the key figure in importing the term *cyberspace* from the world of science-fiction-fan programmers into middlebrow discourse; hereafter, the internet could be envisioned, not just as a tool or set of devices with predictable potentials, but as an unknown space to be explored and thus available for any number of collective projections, particularly the frontier metaphor—made explicit in the name of Barlow's foundation. But this particular refiguring of the frontier metaphor was also heavily inflected with tropes from the 1960s counterculture; Barlow's missive featured a studied informality ("we'd both been up in the same saucer or something," "Mitch," "Bill"); a pleasure in iconoclasm ("if you don't appreciate the irony"); and a flamboyant individualism (in the EFF's relentless focus on personal privacy and liberties). But this bit of computer counterculturalism also had an association with power (the CIA!). And, crucially, in a classic countercultural maneuver, instead of flattering his audience or downplaying his differences from them, Barlow offers them a choice between being one who gets it and one who doesn't. Accept his rhetorical universe, and you are a "native of the future." Reject it, however, and you are threatened with always being an immigrant there.

At the time, reading a missive like this on a monochrome screen, perhaps during a slow day at the office or perhaps late at night at home, had an arresting effect. Barlow's email suggested to the lone cubicle dweller who had mastered email that a new sense of energy was emerging in the online world. The incongruous juxtaposition of a Grateful Dead lyricist with CIA officials was funny, of course, but also enticing; how many people get invitations to talk to CIA officials, much less go on to tweak the officials' noses and get away with it? Here was someone whose tax bracket and espionage experience were probably comparable to yours, yet he was boldly preaching to an established, powerful, and sometimes violent institution. The situation suggested a new opening, a new avenue towards power. As a white-collar reader of this text in early 1993, you felt uniquely privy to this intriguing opening because you were among the elite few who had mastered the arcane art of online access. The relative obscurity of the procedures needed to get the message only added to the aura of being part of a special group. You, who both got the joke and technically could get access to it, were invited to be one of the vanguard, one of Barlow's "natives of the future." It gave you a new sense of what it meant to be sitting in one's office typing, a new, hipper, less ordinary, sense of self.

The effect was indeed delicious.

In the early 1990s, growing numbers of professionals and white-collar workers were being surprised by this kind of experience on their desktop computers. As the number of people with some variety of online access increased from month to month, more and more people had an experience of stumbling upon something striking; it could be a surprising exchange on an email discussion list, involving a tidbit of insider information from afar. Or it could be a titillating personal revelation; this was the moment when stories of email romances began to circulate in popular folklore. It could be a new form of access to something or someone, like the personal MTV gopher created as a hobby by MTV vee-jay Adam Curry; accessing his gopher gave one a kind of personal access to a media figure, to someone ordinarily shielded behind the glossy professionalism of the television screen. (Fans of this gopher were treated to a Barlow-like iconoclastic moment in April 1994 when Curry, with a 1960s flourish of rebellion, announced his resignation from MTV on air. He was resigning in order to pursue his digital activities full time, on the then-astonishing theory that the digital world was the wave of the future, and television was obsolete.)[3] Something out of the ordinary, it seemed, was afoot.

By the early 1990s in the United States, the microcomputer industry had triumphed, but it was no longer enchanting. Microcomputers so far were turning out to be office machines, and most of the computers in the home turned out to be mostly ways to do office work after hours. Most of the 1980s efforts to sell computers specifically for the home market—for example, the Sinclair, the Commodore 64, the Atari, the IBM PCJr—had gradually disappeared. The little computers did not seem quite so personal anymore. Not only had the desktop computer become a commonplace of office life, with all its associations with bureaucracy, but the companies that made microcomputers no longer seemed like the boisterous garage start-ups of popular capitalist mythology; by 1990, Microsoft had pushed aside the grey, arrogant, predictable monopoly of IBM—and replaced it with another grey, arrogant, predictable monopoly.

And, as Barlow's message was circulating in email discussion lists and newsgroups, the first issue of Wired had just hit the newsstands; within a year it would have a circulation of over 100,000 and a curious readership several times that.[4]

From "Information Superhighway" to "Cyberspace" and the *Habitus* of Knowledge Workers

The internet enthusiasms of the 1990s have frequently been described as utopian, but it was really the information superhighway scenario that was utopian, in the sense of offering a blueprint for a better future. Cyberspace, by contrast, was made famous in the precedent-setting novel of cyberpunk fiction, Gibson's *Neuromancer*, which depicts a near-future world of technological violence, cruelty, manipulation, and cynical disaffection—a world that is distinctly dystopian. The appeal of *Neuromancer* is less utopian than romantic. Its tale of a "console cowboy" is a narrative of an outcast hero on a desperate quest initiated by a search, not for wealth, but for inner transformation—for love, comradeship, and meaning, but also, as if to make the inner transformation point vivid, an internal physical reprogramming so as to allow for better net traveling.

To the computer-operating white-collar worker in the early 1990s, *Neuromancer* provided a story line that redefined the act of sitting at a keyboard entering commands from one of white-collar drudgery into an act of exploration and adventure. Cyberspace, by defining the internet as a space, a territory for adventure, rather than as merely a highway, a means towards the end of accessing already organized information, suggested a new potential self-definition for knowledge workers. *Information superhighway* sounds clean, obedient, and orderly. The connotations of *cyberspace* are darker, less regimented, more scary—but thereby more thrilling. Late at night, alone in one's cubicle, *cyberspace* had a much more alluring ring. Cyberspace did not offer a utopia, a perfected world; it offered a taste of rebel-hero selfhood.

We like to think that romances and revolutions come from nowhere, as if they are their own explanation and driving force. But of course there's generally a context like a mid-life crisis or a frustrated and underemployed middle class that sets the conditions for the change. The same is true of the spread of online computing in the early 1990s.

To understand why *cyberspace* outlived *information superhighway* in popular usage, it helps to consider exactly what kinds of people were getting online access in the early 1990s. Typical discussions of social class and computer use focus on a haves versus have-nots continuum, where the concern is extending the benefits of computer use lower down on the class ladder. But it is also illuminating to look upwards on the ladder as well: Both Bill Gates and the janitor that empties your office trash bin can get along fine without desktop computers in the day-to-day of their work lives. Computers have become a central feature of the work lives specifically of the knowledge or professional classes, a group that includes middle managers, engineers, mid-level government bureaucrats, academics, and journalists—white-collar knowledge workers.

Crucial to the character of the early 1990s, then, was the fact that online access came first among those who *did their own word processing* and thus had the necessary equipment and experience readily at hand. Graduate students and assistant professors were online before university presidents and provosts. Middle managers, technicians, and engineers were online before vice presidents and CEOs. Mid-level journalists were online before editors and managers. This is a relatively unusual pattern of technological diffusion; networking entered social life through the same portal as the photocopy machine rather than through the top-down diffusion patterns of the telephone or the consumer-distribution patterns of television. This pattern thus meant that the sense of something important happening in networking would hit the middle ranks of the knowledge class before it hit their superiors.

The information superhighway, and the corporate liberal technology policy for which it stood, may have been reasonable, forward looking, and economically rational. But it lacked a sense of enchantment. Developing government-business partnerships that would encourage investment in wide-area computer networking for purposes of information exchange may have been a good idea, but, for the typical cubicle dweller, it did not generate much fire in the belly.

In the years leading up to 1995, the stage was thus set for the middle ranks to be treated to a drama of obliviousness from above, an object lesson in high-level bewilderment. It was the people who typed their own memos, reports, term papers, and journal articles who sensed the importance of the internet first and then watched the higher-ups struggle to catch up with them. *Cyberspace*, with its romantic hint of a rebellious self image, better captured the sense of pleasure and open-ended possibility they felt in watching their secret world trump the staid world of their superiors.

Tropes from the Counterculture: "They Just Don't Get It"

Louis Rossetto, cofounder of *Wired*, had experienced both the original 1960s counterculture and its emerging computer-culture variant, the former as a college student, the latter as the editor of a small journal about desktop computer publishing, *Electric Word*.[5] He has said he modeled *Wired* on the early *Rolling Stone*—the sincere, preironic, early 1970s *Rolling Stone*, when it was based in San Francisco and celebrated rock stars as oracles of a revolution in human consciousness.[6] Rossetto frequently dismissed mainstream media's technology coverage with the phrase, "they just don't get it."[7] The phrase is part of the rhetorical foundation of outlets like the hacker website Slashdot or *Wired*; in the constant cavalier dismissal of vaguely defined, "old" institutions and points of view (for example, Microsoft, television networks, government bureaucracies, Keynesianism) these media are flattering readers by implicitly including them in the knowledgeable avant-garde. As John Perry Barlow was fond of implying, You are one of us, the mammals, and those powerful people are the dinosaurs.

When a marginal social movement accurately anticipates in the public eye a significant historical failure of judgment on the part of leadership, the effect can be powerful. Being right about something when the powers that be were wrong, for example, was a central collective experience of the 1960s counterculture; by 1969, the world had watched the television networks, the *New York Times*, and many members of the political establishment change their position on the Vietnam War. In the mid-1990s, it would be the failure to anticipate the importance of the internet, and, in the late 1990s the value of open software. And part of the power of such moments is that they open the door to iconoclasm and to new currents of thought; if the authorities are wrong about that one thing, what else might they have missed?

At the same time, this kind of collective experience establishes the conditions for a less clearly beneficial drawing of boundaries between those who knew and those who didn't. What this phrase does is tell the listener that he or she and the speaker are part of the elite group who get it. The ones who don't could be the Pentagon, the media, or your parents; in any case, there's a thrill in the implication that you and I stand apart from despised others in the world.

If being right about some central event like Vietnam or the internet gives the rhetoric of getting it force, accuracy in general is not necessary or even a precondition for the rhetoric to work. The internet was only discussed in passing in *Wired*'s first issue; Rossetto had to catch up to the centrality of the internet like everyone else in the media. And, more importantly, once the rhetorical ground is established by whatever means, a powerful trope for shutting down inquiry is made available. In the interview mentioned above, when Rossetto was asked if he's religious, he replied "no." When asked if he's an atheist, he also replied "no," and then continued: "It's not worth thinking about. . . . I mean, I've gone beyond it."[8] The rhetoric of "they just don't get it" can create conditions that make this kind of shutting down of inquiry sound wise. The reader or hearer is made automatically wary of voicing any criticism, questioning, or complexity, even to themselves. Express doubts, and you risk being worse than wrong, you risk revealing yourself to be a dinosaur and thus no longer part of the privileged club; you just don't get it.

The Moment of Mosaic: The Pleasure of Anticipation

By mid-1993, then, a growing crowd of mid-rank white-collar computer users was quietly gaining access to networked computing; and a growing portion of these were learning about and using the nonprofit, nonproprietary internet. These experiences were becoming increasingly inflected with countercultural habits and iconoclasm, and the higher ranks of leadership, the CEOs and politicians, were largely oblivious to it all. This context proved extremely fertile ground for a new, freely distributed computer program called Mosaic, the first successful graphical web browser. Mosaic 1.0 for the Macintosh and PC was released in August 1993 and spread like wildfire through the fall of that year. The program created an almost instant wow effect, motivating ordinarily bored or preoccupied cubicle dwellers to call a colleague and tell them, "you gotta try this thing."

This was where it all started. This was the moment of take-off in the internet frenzy of the 1990s.

Mosaic, it needs be said, was neither the first web browser nor even the first graphical web browser.[9] When two employees of the University of Illinois's National Center for Supercomputing Applications (NCSA), Eric Beena and undergraduate Marc Andreessen, decided to program a better browser near the end of 1992, they were simply making their own contribution to an ongoing networking software evolution based on ideas that were already very much in the air. Their main technical contribution in the first version of Mosaic for Unix was the ability to display images within the page and a slicker, more inviting interface. Another important contribution was the production of PC and Macintosh versions of the browser released in August 1993. These versions, programmed by a larger team of mostly undergraduate programmers, made the browsing experience more widely available.

Technically speaking, then, Mosaic was a useful but modest contribution, arguably not as important as, say, SMTP, the WWW protocol itself, or the SLIP and PPP protocols that enabled connection to the internet via a modem. And Mosaic was clearly not as important a technical contribution as the underlying TCP/IP packet switching protocol and all the software that had been written to implement it on a wide variety of computers. Mosaic did not make it possible to connect to the internet. Other programs and protocols did that. And Mosaic did not make the internet friendly; it simply made it somewhat friendlier. And it is safe to say that it was not a question of efficiency; *Mosaic* was a slow and cumbersome way to get information, particularly on the graphics-impaired computers of the first years. Mosaic was a fine program, but it was not a revolutionary work of genius by any definition.

So why did Mosaic become the killer app of the internet? Why did its direct successor Netscape launch the "internet economy" of the 1990s? Part of it was simply the cumulative critical mass of people and technologies, what some economists call network effects; enough computers were becoming graphics-capable, enough of those computers were becoming connected to LANs, and enough of those LANs were being connected to the internet, that being on the internet was becoming more and more valuable.

But it's crucial that Mosaic wasn't so much efficient as it was pleasurable; using Mosaic was one of the first really compelling, fun experiences available on the internet. Some computer professionals tried to downplay it for that very reason: "Mostly, people use Mosaic to show off the money they spent on their PCs," observed one software executive, "you can call somebody over and say, 'Look at this.' It has got that kind of whiz-bang appeal. . . . It's like the first time you go through the library: It's fun to wander through the stacks, pulling down books. But that does wear off."[10] But of course we now know in retrospect that the fun of web browsing was not about to wear off any time soon.

What kind of pleasure did Mosaic offer? Mosaic did not satisfy desire, it provoked it. Colin Campbell has described what he calls "modern autonomous imaginative hedonism," a distinctly modern structure of pleasure in which the anticipation of pleasure becomes part of the pleasure itself and which is characteristic of the consumer culture and romanticism generally.[11] What one wants in this peculiarly modern form of pleasure, Campbell argues, is not the satiation of desire but desire itself; it is the desire to desire. Mosaic did not so much show someone something they wanted or needed to see as it stimulated one to imagine what one *might* see. One of the early classic ways to demonstrate the web was to click onto the website for the Louvre, to watch grainy images of paintings slowly appear on the screen. This was not pleasurable so much in what it actually delivered—better versions of the same images generally could be found in any number of art books—but in how the experience inspired the viewer to imagine *what else might be* delivered. Mosaic enacted a kind of hope; it did not deliver new things so much as a sense of the *possibility* of new things. Surfing the web using Mosaic in the early days shared certain features with the early stages of a romantic affair or the first phases of a revolutionary movement; pointing, clicking, and watching images slowly appear generated a sense of anticipation, of possibility. To engage in the dreamlike, compulsive quality of web surfing in the early days was an immersion in an endless what's next?

The Genesis of Irrational Exuberance: Romanticizing the Market

By May 1993, white-collar workers scattered in cubicles and offices across the land were quietly discovering the thrills of going online, as Andreessen and others worked on the code for Mosaic, as the likes of John Perry Barlow and the editorial staff of *Wired* were spreading the tropes of the computer counterculture to the middle ranks. But the mainstream was still thinking other things. That month, *U.S. News and World Report* published a "technology report" about the coming future of networked computing. The article, clearly a response to the enthusiasm surrounding Vice President Al Gore's information superhighway initiatives, had no mention of the internet. It began,

> The melding of the telephone, television and personal computer today has unleashed a dynamic digital revolution that promises to radically alter the way people live, work and play around the world. What new products and services can we expect from this technological upheaval? How big a market, exactly, are we talking about? And what, if anything, should the Clinton administration do to help foster these emerging technologies in America?[12]

This was the conventional way of understanding things at the time—in the business terminology of products, services, and markets. From there the article went on to seek answers from "seven titans of technology": Bill Gates, shopping-channel pioneer Barry Diller, AT&T Chair Robert Allen, cable TV tycoon John Malone, IBM board vice chairman Jack Kuehler, cell phone magnate Craig McCaw, and Motorola chair George Fisher. The article was thus organized around the assumption that, whatever happened, it would be shaped primarily by corporate leadership and corporate concerns, perhaps in interactions with government regulators spurred on by initiatives coming from the White House. Gates predicted a wallet-sized personal PC interconnected with home appliances. Others forecast a lucrative cornucopia of online shopping, ubiquitous multimedia communication for business executives, movies-on-demand, distance education via cable TV, and growing wireless data services. All expressed an ambivalence about government's role, expressing appreciation for the excitement generated by Gore and the Clinton White House but cautioning government regulators to stay out of the way of corporate initiatives. This was a view from the top.

Three months later, as Andreessen and his colleagues were quietly releasing the first version of Mosaic for the Macintosh and the PC, the August issue of *Scientific American* appeared with a similar overview article.[13] In keeping with its more sophisticated readership, the article contained much more technical detail, comparing the bandwidth and cost of various transmission technologies like fiber optic cable and ISDN, for example, and its interviewees were generally further from the boardroom and closer to the research lab. But the basic organizing assumptions of the article were the same as the recent *US News* piece; the bulk of the article focused on various corporations and their technologically linked interests, comparing and contrasting the schemes of various cable, phone, and media companies, interspersed with various inside-the-beltway regulatory concerns, such as common carrier principles. The list of possible applications of the coming technology leaned a little more towards information and education than shopping: a cut-out box contained descriptions of school children communicating by email with a researcher in Antarctica, a broadcast engineer who helped diagnose his daughter's rare disease using online databases and email, an experiment with sending dental X-rays across the Atlantic, and a group of New Jersey schoolchildren communicating with teachers in Russia.[14] The World Wide Web was not mentioned.

This article, however, *does* mention the internet. It opens with an anecdote about Internet Society President Vinton Cerf preparing for a Congressional hearing by contacting thousands of enthusiasts over the internet, pointing to the rapidly growing activity on the internet as a potential "seed" of Gore's "National Information Infrastructure."[15] And the article is titled "Domesticating Cyberspace" and closes with a Congressional Representative echoing Barlow's metaphorical construction of the online world as a frontier: "Anything is a danger in cyberspace.... There are no rules. It's the Wild West."[16] The Barlow-inspired metaphorical constructs of its title and opening and closing paragraphs—a vision of a wild, expansive, exciting space in the internet—would prove to resonate more profoundly than the content about corporate struggles, educational applications, and competing delivery technologies. Perhaps *Scientific American*'s readership is specialized, but that readership includes politicians, executives, and, most importantly at this moment in history, reporters—reporters being among the category of people who do their own word processing.

This was the moment that the internet hit the media radar. As summer turned to fall in 1993, the internet rather suddenly became an object of media fascination. Scott Bradner, a long-time internet insider,[17] observed with some bewilderment that "the Internet is suddenly popular.... For reasons best known to the media gods, articles about the Internet seem to be the thing to do these days." He pointed out that during the fall of 1993, a time when only a miniscule number of people actually had internet access, 170 articles appeared in major U.S. publications mentioning the internet, as compared to 22 articles in the same period a year before.[18] He continued,

> All this attention is flattering to those of us who have been proselytizing this technology for years. The problem is that I don't see any logical reason for the current attention. The Internet has been around and growing for more than a decade. Sure, it's big (almost 2 million interconnected computers world wide) and growing fast (more than 7% a month), but it's been big and growing fast for quite a while now. It was certainly growing at least at this rate when *Time & Newsweek* were forecasting national video parlors for the kiddies instead of international on-line, real time, interactive current affairs in the schools.... Last month, I even found an article on the Internet in an airline flight magazine.[19]

As the excitement around the internet gathered in the media, as *Wired* and John Perry Barlow framed computer networking in breathless countercultural terms, as Mosaic circulated onto the increasing number of internet-connected LANs, members of the business world began to take note. In part, with their attention already directed towards networking by the information superhighway rhetoric, it might be unsurprising that those looking for business opportunities might follow the media towards the internet. But another key factor was the Microsoft monopoly, which played a dual role. On the one hand, Microsoft's dominance in operating systems represented the uninspiring end of the garage start-up days in microcomputing, thus motivating romantic entrepreneurs to look for something new. On the other, it was well known that Bill Gates had just become one of the richest men in the world and that those who had heavily invested in Microsoft in the late 1980s were beginning to reap fabulous rewards. Microsoft was thus both a reviled corporate monolith and an object lesson: might something overthrow Microsoft just as Microsoft had overthrown IBM? Might that something have to do with the internet? And might there be similar rewards to be reaped by those who accurately guessed what the next best thing would be?

Enter Netscape. Jim Clark, student of Lynn Conway and founder of graphics workstation company Silicon Graphics (SGI), was by that time an executive who did not need to do his own word processing. But sometime in the late fall of 1993, just as the internet craze and Mosaic were entering mainstream attention, Clark stumbled upon Mosaic when, in search of a new direction in technology, he was introduced to the web browser by an underling at SGI. Clark resigned from SGI in February 1994, flew to the NCSA in Champaign-Urbana, Illinois, found Marc Andreessen, and founded a company to commercialize the program in spring 1994.[20] With unprecedented haste, he launched the Netscape IPO just over a year later. This became the most successful IPO in history to that point and the model for many subsequent IPOs, setting off the internet stock craze. The largest party that capitalism has ever thrown had begun.

So why did Netscape get all this swooning attention? In part, Netscape grabbed the headlines because it was in Silicon Valley, in part because of Jim Clark's previous track record with SGI, and in part because Netscape hired Andreessen and many of the other original programmers of Mosaic. And Netscape's first browser was a good one, particularly as it channeled its start-up funds heavily into rapidly improving the program, releasing frequent free updates over the internet and quickly becoming the most popular browser. Yet it needs to be remembered that at the time of the IPO the company had no profit and almost no revenues. It was giving its principle product away for free and had no crucial patents or other dramatic advantage in the browser market: Netscape was just one of about ten companies trying to commercialize Mosaic.[21]

To a very significant degree, Netscape gained so much attention because it followed a deliberate strategy of creating a media narrative heavily centered on a romanticized, heroic construction of the computer counterculture, which proved very popular with the media itself. Netscape depicted itself as enchanting. Very early on, Clark hired a publicist, Rosanne Siino from SGI, and told her to present Andreessen as the rock star of the company.[22] Siino then developed a strategy that carefully cultivated media attention framed in terms of geek chic, deliberately taking reporters into the back rooms to show the chaos of the programmers' cubicles, programmers sleeping under their desks, and so forth.[23] And she successfully turned Andreessen into a celebrity; in 1995, Andreessen appeared on the cover of *Forbes ASAP* with the blurb: "This Kid Can Topple Bill Gates."[24] Andreessen would soon be featured in *People* magazine and appear on the cover of *Time* in his bare feet.[25]

Arguably, none of this would have worked without *Wired* magazine. *Wired* was barely more than a year old when Clark hired Andreessen, and in that year it had been full of adolescent hyperbole (Rossetto claimed that computer technology was creating "social changes so profound their only parallel is probably the discovery of fire"),[26] and inaccurate predictions (besides mentioning the internet only in passing in the first issue, the second implied that Richard Stallman's Free Software project was outmoded and doomed).[27] And its eye-catching Day-Glo graphics and layout were sometimes unreadable. *Wired* did not, furthermore, invent the idea that Mosaic was the killer app of the internet.[28]

But it popularized the idea and did so in a particular way. *Wired* published its first substantial piece on the Mosaic phenomenon in October 1994, when the web was well known to internet aficionados but just beginning to attract the attention of the wider world.[29] Titled "The (Second Phase of the) Revolution Has Begun," the article, by Gary Wolf, didn't just target a good investment or new technology as a normal trade magazine piece might have. Instead, it used colloquial language and emphasized revolutionary change, pleasure, and personal expression. "When it comes to smashing a paradigm," the article began, "pleasure is not the most important thing. It is the only thing." In a section titled, "Why I Dig Mosaic," Wolf observed that "Mosaic functions lurchingly, with many gasps and wheezes," and described an experience of setting off to find technical information on the nascent web and getting distracted by the process of clicking from link to link, eventually ending up on a physicist's personal page. But while a standard review of computer software might point to these as problems, Wolf did not. "The whole experience," Wolf wrote, "gave an intense illusion, not of information, but of personality." Now that personal computers had revealed themselves to be just office machines and therefore not all that personal, Wolf was locating personality in the act of web surfing.

And, as if to give substance to the rather thin idea of the personality of the web, Wolf crafts an image of the personality of Mark Andreessen. Another journalist might have interviewed the senior partner of the company, but Wolf focused on the twenty-something junior partner and notes personal details. "A little way into the interview," Wolf notes, "Andreessen removes his dress shirt and answers the rest of my questions in a white T-shirt. This gesture leaves the impression of a man doing battle against the businesslike backdrop" of Netscape's headquarters. And Wolf focuses on Andreessen's dragon-slaying attitude; working for Netscape, Wolf notes, "offers [Andreessen] a chance to keep him free from the grip of a company he sees as one of the forces of darkness—Microsoft."

By zeroing in on Netscape and Andreessen, this *Wired* profile, not only amplified the belief that *Mosaic* was the killer app of the internet and that Netscape would be its primary beneficiary, but also offered a romantic lens through which to see the phenomenon. As the Clinton White House and Congress took the information superhighway rhetoric off into dry committees, spouting inside-the-beltway acronyms, businessmen could increasingly be seen thumbing through copies of *Wired* on airplanes, and terms like *cyberspace* and frontier metaphors began cropping up in newspaper articles and politicians' sound bites. Without *Wired*, it's not obvious that this libertarian-flavored countercultural framing of computer networking would have taken hold in the mainstream.

In the ensuing years, as we all know, take hold it did. Being part of the knowledgeable vanguard was a central part of the ethos at the time. Mary Meeker, who later would be dubbed "the girl in the bubble," was a stock analyst at Morgan Stanley and key player in the Netscape IPO and many subsequent dot com IPOs. She said, "I remember that in 1995 I would speak with Marc Andreessen and we would try to count up how many people understood this stuff. We thought it was about four hundred."[30] Soon after the Netscape IPO, a young visionary called Jeffrey Skilling began leading a rising corporation called Enron into new territory based on the speculative trading of energy and internet-related activities; of skeptics of his strategy, Skilling is reported to have scoffed, Rossetto-like, "there were two kinds of people in the world: those who got it and those who didn't."[31] And it left its mark in politics; in the summer of 1994, conservative pundit George Gilder teamed up with futurologist Alvin Toffler and others to release a rousing document called "Cyberspace and the American Dream: A Magna Carta for the Knowledge Age"; the document declared a new era in which free markets and technology would make governments obsolete, a set of themes that would soon be picked up by then-Congressman Newt Gingrich.

After a point, it does little good to scoff at these patterns with a smug "I told you so." Many thoughtful observers knew the stock prices of the late 1990s were irrational, and many of them said so. The evidence and arguments were there, but the bubble kept expanding nonetheless. It is worth ferreting out those who took advantage of the heady atmosphere to engage in various degrees of outright fraud. But one cannot blame the heady atmosphere itself only on occasional instances of exaggerated reporting, conflicts of interest, or dishonesty.

This overheated atmosphere was precisely a fusion of the desire for wealth with romantic dreams of freedom, self-expression, and the dramatic overthrow of the powers that be. Without the romantic visions of freedom and revolution, there would have been nothing to get excited about; there was no gold in this gold rush, no valuable raw material, just castles in the air made of projections onto immaterial digital bits; something had to make those projections seem valuable. Yet without the hope of getting rich, the enthusiasm would never have had the energy it needed to spread. Change the world, overthrow hierarchy, express yourself, *and* get rich; it was precisely the heady mix of all of these hopes that had such a galvanizing effect.

The Role of Romanticism in the Internet Surprise

The development and character of capitalism sometimes tends to be imagined in terms of things in the guns, germs, and steel category, things like technologies, resources, and geography. Economic development, we like to assume, is about the efficiency with which the prosaic fundamentals of human life are produced and distributed or about things that change how much we have to eat or how long we live. Yet this way of thinking, for all its insight, can make us forget just how frivolous the development of capitalism has been at times. Think of the role of gold, spices, tea, tobacco, and beaver pelts in the development of mercantile systems and the early phases of the European colonization of the Americas. These were all trivial commodities, of no major practical value, whose popularity largely reflected the whims of European upper-class fashion. Yet the basic systems of accounting and trade that laid the foundations for early modern capitalism, some would say for capitalism and the world system itself, were created around them.

The internet of the 1990s may have been less like the steam engine or the radio and more like spices and beaver pelts. What the internet offered, however, was not so much fashions for decorating our bodies or our food as fashions for clothing the self.

In a few short years, between 1992 and 1996, the internet went from being a quiet experiment to a global institution whose name seemed to be on everyone's lips and whose existence and importance was taken for granted. By 1995, the remaining consumer computer communication systems from the 1980s like Compuserve and Prodigy were all selling themselves as means of access to the internet rather than the other way around, the Congress was revising the structure of its communications law for the first time in more than half a century, major corporations from the phone companies to Microsoft to the television networks were radically revamping core strategies, television ads for Coke and Pepsi routinely displayed URLs, and the stock bubble was underway.

In the history of media, this is an extraordinarily rapid shift. In comparison, the early histories of most other media suggest something well coordinated and planned. For example, the general outlines of the TV industry in the United States—the major corporate players, the advertising system, networks providing programming to affiliate stations, even much of the programming like soap operas and variety shows—were clearly mapped out by the mid- to late-1930s, more than a decade before its full-scale introduction around 1950. (RCA/NBC made television a central part of its plans for the future by 1932, just as the networked/advertising-supported radio broadcast system was becoming consolidated.)[32] For all the complexities and struggles associated with their development, film, television, and VCRs were disseminated in a context in which industry leaders, government regulators, and manufacturers all shared similar broad general outlines of what the new industries were going to be about, and disputes were limited to fine points of technical standards (for example, Betamax versus VHS), revenue distribution (for example, cable and videotape copyright issues), and the like. Historians of technology are correct to scoff at hyperbolic claims about the internet (for example, that the internet is the biggest invention since the book). But one of its striking features is just how much, for a period in the 1990s, it took people by surprise. Arguably, the only communication system of the twentieth century that came with similar unexpectedness was radio broadcasting (which also played a significant role in that other stock bubble in the 1920s).[33]

Many might ask: Did not the internet triumph because it offered new efficiencies, new flexibilities, and a new ease of access to information and knowledge creation, which together were really the driving force of the internet itself and its contribution to the new market dynamism, globalization, and the flattening of economic relations? Could that much money and effort really be driven by something as ephemeral and irrational as a particular kind of romantic self-concept?

Other factors mattered, of course. But it seems likely that the romantic entrepreneurial self-concept was a necessary if not sufficient component of the internet explosion of the 1990s. It helps to remember the order of events. Everything changed *before* significant numbers of individuals were successfully making a profit selling via the internet and even before internet access was widely available. It was not as if people started using the internet to make money and then the business world at large responded; in 1995, there *was* no appreciable nonmetaphorical market yet, no substantial population of individuals competitively buying and selling things over the internet. There were, by and large, only experimenters, speculators, and enthusiasts, people who expected a market to emerge where one did not yet exist. (Yes, there were those graphs showing skyrocketing numbers of internet nodes and users, forming almost perfect triangles as the lines headed for the upper right corner. But one could produce similar graphs for the sales of the latest new pop rock sensation, and the internet had been growing at a similar rate long before 1993.) Understood as a practical technology, the internet could not have caused these changes because the internet was not practical yet. Many of the more important changes that set the stage for the explosion of the internet and the internet economy, then, happened *before* the technology itself had a chance to have much concrete effect at all.

The fact is, the internet that appeared in the 1993–95 period wasn't just a technology; it was the enactment of a hope. The changes of 1993–95 were very much *anticipatory*, changes based on what people *imagined could happen*, not what had already happened. In the early 1990s, the internet did not so much cause new things to happen as it served to inspire people to *imagine* that new things would happen. The shared embodied experience of an immersion in an endless what's next? that Mosaic unleashed, coupled to the enormous awe accorded new technology and the opportunity to step into the role of the romantic entrepreneur, enabled new behaviors on the part of significant segments of the population. People behaved differently, not just because the internet enabled them to do different things, but because its presence inspired people to imagine how things *might* change, how things might be transformed. Many of the things said and done in the name of the internet in the 1990s we now know to be misjudgments, some of them colossal ones; those misjudgments, however, were not random. They were part of a pattern of shared collective vision, and that vision had an impact even if it was based on some shaky foundations.

The Internet and the Revival of Neoliberalism

That impact, moreover, was not fleeting. The internet enthusiasm was not just a Tulip mania, a mediocre idea that eventually washed away in lost fortunes. On the one hand, the internet bubble unleashed a flood of money into the telecommunications infrastructure, effectively making TCP/IP the standard for the foreseeable future and resulting in a build-out of high-bandwidth lines that remained important even after the bubble burst. On the other hand, less concretely but perhaps more profoundly, the internet surprise helped revived neoliberalism. As we have noted, neoliberalism looked like it might be in serious decline at the time of Bill Clinton's election in 1992; both corporate leadership and government were moving back towards a more classic corporate liberal point of view involving coordinated private/public partnerships. But somewhere between the collapse of the Soviet Union and the internet bubble, the blind faith in markets was given new energy. The internet, because it took the establishment by surprise, because it occasioned a confluence of the counterculture with spectacularly speculative capitalism, created an opening for big ideas.

Given the political alignments of the early 1990s, the first intellectual bloc to jump into that opening were the libertarians, and they successfully fostered an equation of the romantic individualist vision of the internet with a free market vision of the internet. Esther Dyson—a business consultant with a background in journalism, a door-opening last name, and libertarian leanings (she told *Reason* magazine that what she likes about the market economy is "number one, that it works. Number two, that it's moral")[34]—noticed the emerging internet as an opportunity and made it the core of her career. Republican speech writer George Gilder, previously famous for books attacking feminism and welfare, turned to technology with a book on the end of television in 1990; the book proved not to be all that accurate in its predictions but its enthusiastic reception at the time taught Gilder the power of technological prognostication as a route to an audience. Libertarian journalist Declan McCullough started an online website devoted to the internet and politics in 1994 and soon became a regular contributor to *Wired*. This group was able to take the framework laid out by de Sola Pool's *Technologies of Freedom*, in which free markets are equated with free speech and government regulation is construed as the enemy of both, and apply it to the exploding activity around the internet.

To see how this feat of rhetorical association worked, consider a 1995 essay about the internet in *The Economist*. Author (and eventually, *Wired* editor) Christopher Anderson wrote, "the Internet revolution has challenged the corporate-titan model of the information superhighway. The growth of the Net is not a fluke or a fad, but the consequence of unleashing the power of individual creativity. If it were an economy, it would be the triumph of the free market over central planning. In music, jazz over Bach. Democracy over dictatorship."[35] This vision of individual creativity unleashed (and the associated rhetorical flourishes) was orthodox romantic individualism. But this was in *The Economist*, not *Wired*, and what readers of the magazine first saw when they opened the issue added a telling twist. A nameless editor at *The Economist* compressed Anderson's romantic passage into a pull quote that graced the first page of the printed version of the article: "The explosive growth of the internet is not just a fad or a fluke, but the result of a digital free market unleashed."[36] Readers of *The Economist* were subtly invited to leap from "if it were an economy" to it *is* a "digital free market." This associative leap from market metaphor to actual market, repeated over and over again in the culture at the time, allowed a technological system based on nonproprietary standards and with roots in a mix of private and government funding to become a powerful archetype of both the free market and free expression in the broad public mind. The iconoclastic rhythms by which the internet appeared in the cubicles of America were thus harnessed to a Lockean market projection. The illogical leaps inherent in all this were swept aside by the speed of events and by the constant threat of being labeled as a dinosaur by *Wired*-style countercultural rhetoric.

These events, occurring in the wake of the astonishing collapse of the Soviet Union and the quick triumph of the United States in the 1991 Persian Gulf war, created a context in which neoliberalism was given new life. In law, politics, and business management, most of the talk of industrial policy and related corporate liberal terms disappeared into the background, and the faith in markets as the source of all innovation (and of freedom itself) was revived as forcefully as ever. In 1994, just as Netscape was plotting its IPO, the *New York Times* transferred a Middle East foreign correspondent named Thomas Friedman to covering the White House and economics. The following year, as the stock bubble began its climb, Friedman became a regular op-ed writer, and he cultivated an avuncular, anecdotally driven, free market fundamentalism that over time he coupled to a subtle nationalist triumphalism. In the face of all this, those who initially raised doubts about specific neoliberal decisions, such as economist Jeffrey Sachs's belief in shock therapy for the former Soviet Union to drive it into a market economy—with results that even Sachs admits were mixed, and many would argue were disastrous—were largely rendered mute throughout the 1990s, cowed by the apparent irrefutability of the market vision, buoyed as it was by the flood of events, the energy around the internet, and the associated meteoric rise of the stock markets. Thanks to the way the internet was embraced by the culture, Margaret Thatcher's 1980s claim that "there is no alternative" to neoliberalism on the world stage was given new life.

6 Open Source, the Expressive Programmer, and the Problem of Property

> Every good work of software starts by scratching a developer's personal itch.
>
> —Eric Raymond[1]

WHAT IT WOULD take to finally put some cracks in the foundation of the neoliberal consensus, it turns out, was the same thing that gave it renewed life in the early 1990s: romantic individualist representations of computing.

Long before the spread of the internet, even before the appearance of microcomputers, Ted Nelson, in *Computer Lib*, briefly reflected on the problem of funding his system of hyperlinked digital texts he called Xanadu:

> Can it be done? I dunno. . . . My assumption is that the way to this is not through big business (since all these corporations see is other corporations); not through government (hypertext is not committee-oriented, but individualistic—and grants can only be gotten through sesquipedalian and obfuscatory pompizzazz); but through the byways of the private enterprise system. I think the same spirit that gave us McDonald's and kandy kolor hot rod accessories may pull us through here.[2]

Though little noted at the time, Ted Nelson thus imagined an entrepreneurial form for his digital utopia, in some ways anticipating the neoliberal framing of computing (discussed in chapter 3) that appeared a decade later.

But Nelson was not proposing simply a market for a kind of machine in a box, like the microcomputer. As we've seen, the microcomputer could be easily imagined as a discrete object one buys and sells. How was one to implement something entrepreneurial, a farmer's-market-like system of exchange, out of the vast gossamer web of social and technical links and protocols that is an advanced computer network?

Nelson had an answer. He insisted that Xanadu, while offering a world of hyperlinked texts, also "must guarantee that the owner of any information will be paid their chosen royalties on any portions of their documents, no matter how small, whenever they are most used."[3] Why? "You publish something, anyone can use it, you always get a royalty automatically. Fair."[4] And he argues that this economic fairness, moreover, is of a piece with intellectual fairness: "You can create new published documents out of old ones indefinitely, making whatever changes seem appropriate—without damaging the originals. This means a whole new pluralistic publishing form. If anything which is already published can be included in anything newly published, any new viewpoint can be fairly presented."[5] Nelson is not just a believer in digital property; he hopes that digitalization can *perfect* property.

This chapter explores the emergence, in the second half of the 1990s, of what can be called the problem of property on the internet. This was the period when Linux, the open source movement, and music downloading raised both excitement and consternation in many legal and management circles. By pitting free communication against property rights, these developments called into question the premises of the market fundamentalism that had been driving most political economic thinking associated with the internet to that point. All of a sudden, freedom and the market were no longer synonymous and, in fact, seemed like they might, in some cases, be opposed.

The argument of this chapter is that the internet did not just create *new* problems for intellectual property. It brought slumbering dilemmas with property in general to the surface. In the first instance, this resurfacing of the problem of property was enabled, not by a critique of property per se, but by the tensions between romantic and utilitarian constructions of the individual. The desires to make a profit and express oneself, which as we have seen had been conflated in the early 1990s, suddenly came to point in divergent directions.

"Clean Arrangements": The Dream of Property Rights

Much of the appeal of Ted Nelson's attachment to intellectual property was that it is embedded in a moral vision, not just a dry business model or an economic theory. Nelson, in *Computer Lib*, was clearly not just cooking up a justification for something that would help him get rich. Nelson saw intellectual property protection as of a piece with his idea of freedom. He imagined the computer user as an autonomous, free individual who communicates without the mediation of publishers, libraries, or educational institutions. Digitally enabled intellectual property protection, he believed, would empower that kind of individualism.

Someone like Richard Stallman might argue that such proprietary computing introduces constraints, barriers, and lawyerly and managerial meddling. Nelson's response is this: "I've heard . . . arguments, like 'Copyright means getting the lawyers involved.' This has it approximately backwards. The law is ALWAYS involved; it is CLEAN ARRANGEMENTS of law that keep the lawyers away. . . . If the rights are clear and exact, they are less likely to get stepped on, and it takes less to straighten matters out if they are. Believe it or not, lawyers LIKE clean arrangements. 'Hard cases make bad law,' goes the saying."[6] This is precisely where Nelson combines computers with Lockean liberalism; the machines, with their enormous capacity for fine calculation, will provide the "clean arrangements" that enable friction-free property relations. With Xanadu, according to Nelson, each individual contribution to the system would be perfectly preserved and perfectly rewarded; the computer system itself is supposed to prevent the possibility of unattributed theft of ideas because each "quotation" is preserved by an unalterable link that, not only allows readers to instantly call up intellectual sources, but also ensures direct payment for each use.

Nelson's hope that "clean arrangements" (that is, precisely defined property boundaries) provide the key to freedom has deep roots in American tradition, not to mention a long intellectual pedigree. The triumvirate rights of "life, liberty, and property," made famous by John Locke and Adam Smith appeared in various forms in the Declaration of Independence, Constitution, and the Bill of Rights.[7] The phrase, it should be emphasized, puts property rights on par with the prohibition against murder (the right to life) and the claim to freedom itself (liberty). Like these other primordial rights, moreover, property was said to be natural, inherent, something you had by virtue of being born. Early U.S. leadership thereby wrote the right to property into the soul of American society, long before the polity seriously considered, say, universal suffrage.

Protecting that right to property, making sure that what's yours is yours and what's mine is mine, is not just a philosophical position. By most reckonings it is fundamental to the sense of justice in a capitalist world. It is a powerful bit of sociocultural sense making that helps it seem natural and right for people to pursue the ownership of things, even if this results in some people owning a lot more than others. And it feeds into our sense of what law and justice is supposed to be in the first place. Fair, clear rules—the rule of law, not of men, as the saying goes—are obtained by maintaining clear boundaries, boundaries between property, between individuals. Find the clear lines, make sure no one is stepping over them and is yet free to do what they want within in their own property boundaries, and you have justice. As Ayn Rand, America's twentieth-century pop philosopher of property rights, had her fictional hero John Galt say, "Just as man can't exist without his body, so no rights can exist without the right to translate one's rights into reality, to think, to work and keep the results, which means: the right of property."[8] Property, in her view, is not just some legal technicality. The right "to think, to work," is of a piece with the right to "keep the results."

This view does not just appeal to the already rich. By standard measures, Nelson's career has been a checkered one on the margins of the same commercial and educational computing communities that have been so deeply influenced by his ideas. With that in mind, there's something poignant about his vision; it's the vision of an outsider, never entirely secure or well-rewarded by institutions, who has never been treated all that "fairly," who imagines a utopia in which those unfair institutions are supplanted altogether by communities of free individuals working at computer consoles. It's a utopia where there are no arbitrary powers like a corporate monopoly or arbitrarily powerful authorities with careers built on glad-handing or hot air; a utopia where no smug, tenured journal editors can prevent one's article from reaching publication, and no short-sighted corporate executive can arbitrarily deep-six a beloved project on behalf of cost cutting. Nor can any of these people claim an underling's idea as their own. Nelson is proposing to make real a very American ideal: a vision of a mathematically perfect property system, of crystalline rules that finally make manifest "the rule of law, not of men"—enabled, in this case, by computer technology.

As the internet triumphantly spread the habits of Nelson's hypertext into American life in the mid-1990s, however, things would turn out to be anything but crystalline in the realm of property.

Crystals Turned to Mud: The Problem with Property

A look at the history of Western law shows that it is not just new technologies that render property boundaries confusing. Over the last few centuries, the dream of rights that are "clear and exact" has proven to be elusive across many domains. Property relations themselves, of course, have hardly been in eclipse; as capitalism has expanded over the past few centuries, pressure to extend property relations to more aspects of life has grown unabated. With the important exception of slavery, almost no category of things that has been turned into property has ever been turned back into nonproperty. The problem of property, though, is that, as its scope expanded, the character of property grew ever muddier, ever less "clear and exact." Property rights in practice, it turns out, have hardly been the crystalline system that Locke suggested and that Nelson hopes for. Gnawing away at the entire idea of property, then, is a sense that making it work as advertised might be impossible.

It's not just the distortions caused by side effects like the unequal distribution of wealth or that the judicial and political systems keep finding reasons to blur property boundaries with regulatory efforts like zoning laws and environmental regulation. It is also that crisply defined rights on paper appear much less than crisp when one tries to map them onto the real world of human activities. Even one of the most archetypal forms of property—land—seems to turn up intractable quandaries, quandaries which surfaced in legal cases going back to the nineteenth century. Where, exactly, is the line where enjoyment of one's own property stops and interference with another's begins? What if, say, raising pigs muddies the neighbor's streams or erecting a building casts a shadow on a neighbor's garden?[29] Over the years, the more that the finest legal minds applied themselves to these questions, the more possible answers there seemed to be, with the result that the case law in aggregate seemed to grow ad hoc and murky over time.

So, as the twentieth century progressed, the more aspects of life that property relations were applied to the less sense those relations seemed to make. People have been buying and selling things that look and behave ever less like the property that Locke had in mind. In the twentieth century, licenses to drive a taxi in New York City or licenses to broadcast on a particular radio frequency were bought and sold for huge sums, but these things were really not things at all; they were something quite obviously defined and created by the actions of various government agencies.[10] (U.S. law in fact states that a license to broadcast allows the holder "the use of such channels, but not the ownership thereof.")[11] Ownership of a stock does not grant one anything physical; having five percent of the stock in a company does not grant one the right to walk off with five percent of the factory. Stocks, on close inspection, appear less like property and more like an odd and shifting set of entitlements even as they have become a core form of property ownership in the capitalist system.

A famous review of historical variations in American legal approaches to property distinguished between crystals and mud, between legal decision making based on firm, bright-line rules and blurrier and flexible standards.[12] While there has been a certain amount of back and forth between crystalline and muddy interpretations of the law, legal theorist William Fisher has pointed out that the dominant trend for at least the twentieth century has been towards mud.[13] The historical record, in sum, would suggest that crisply defined laws on paper may not be capable of producing a crisply defined system of justice in reality; they are crisp on paper only.

Some philosophers anticipated aspects of this problem nearly from the beginning, noting that the idea of a natural right is a frail one. Jeremy Bentham, for example, when responding to theorists of natural rights, acerbically observed,

> A reason exists for wishing that there were such things as rights. But reasons for wishing there were such things as rights, are not rights. . . . Natural rights is simple nonsense: natural and imprescriptible rights, rhetorical nonsense—nonsense upon stilts.[14]

Bentham's point was that rights only exist when a government takes action to make them exist; without the hand of some government body determining what rights should exist, what form they should take, and how they should be enforced, there is nothing. Rights cannot be the ultimate protection against government action against individuals simply because they *are* government action. In the long run, this line of reasoning would suggest, rights are indistinguishable from a government privilege. Property is not a right that protects us from government. Property rights, like other rights, it would seem, are a *creation* of government.

Bentham's response to this problem was to invent utilitarianism. Rights are not natural, but human self-interest is, he surmised, and the pursuit of self-interest can and should be organized to maximize the happiness of the most individuals. This notion in turn laid the foundation for today's overlapping traditions of neoclassical economics and rational choice theory and has seeped into popular consciousness in myriad ways, such as the contemporary habit of using terms like *incentivize*. This was the logic at work when, in 1976, the twenty-year-old Bill Gates famously complained to the Homebrew Computer Club about members who freely shared the Basic software that was his company's first product; doing so, Gates griped, would make it impossible for him or others to keep writing more software.[15] In contrast to Ted Nelson, Gates talked, not of fairness, but of incentives to make software. Software should be protected because free copying would discourage the creation of more software.

But Bentham's was just the beginning of a proliferation of various intellectual responses to the desire for a system of property rights that could rest on something other than arbitrary state action. Besides Lockean and utilitarian strains of thought, Kant and Hegel both contributed justifications for property rights based on the idea of a respect for personhood or some form of the transcendental subject, and there is a broad range of functionalist but nonutilitarian theories of property rights structured on behalf of one or another version of the social good.[16]

Beyond Property Rights: The Author and the Machine

Each of these theories continues to have its defenders. But, by the late 1970s, after two centuries of experience with theoretical property crystals repeatedly turning to empirical mud, it began to look wise in some circles to declare that property was nothing but a bundle of rights or, even, as law professor Thomas Grey suggested, that property was simply disintegrating.[17] Many academic schools of thought over the years had quietly expressed versions of this argument: the legal realists, some legal historians, some sociologists of law. The Critical Legal Studies movement—which appeared in the legal academy with a bit of storm and fury in the early 1980s—was for a while simply the best known and loudest proponent of the claim that things like property law had for the most part failed of their own accord. Legal rules, at least in the hard cases of the type that make up Supreme Court jurisprudence and thereby gain so much attention, are indeterminate, that is, logically interpretable any number of ways; what shapes ultimate legal outcomes is social context, culture, fashion, ideology—in short, people or, statistically speaking, men. This position had a strong logic to it and had a certain kind of intellectual daring, a sense of staring down difficult truths; while ignored by the vast majority of practicing judges, Critical Legal Studies gained a foothold in law schools. By the 1980s, then, there were several sophisticated and well-established schools of thought from whose point of view Nelson's dream of crystalline property—not just in computer networks, but crystalline property of any sort—seemed like a naive fable.

For historians and sociologists, moreover, the iconoclastic idea that property was just a bundle of state-created privileges raised interesting questions. If property rights were not perfunctory, if they were arbitrary constructions of the state, then exactly *how* did they get constructed, and why did we continue to talk about them as rights at all? Bernard Edelman's *Ownership of the Image* was one of the first works to tackle this question in the realm of communication technology, using the example of the late nineteenth-century encounter with the spread of photography.[18] In a world where copyright law was generally justified by reference to the labor and creativity of the artist—a writer or a painter deserved to own his or her work because of the labor and inspiration he or she put into it—photography introduced a series of quandaries. Was clicking the shutter of a camera really a kind of creative labor deserving protection, or was it merely a minor technical act, like switching on a light? Did the subject of a photograph have some kind of right to the image? Wasn't it, after all, that person's actual face and appearance that shaped the image on the film, not an artist's interpretation of that face? If one believed, like Locke, that the rights were out there somewhere in nature and to locate them was the problem, or if one was an intellectual descendant of Bentham and assumed that one could somehow scientifically determine a social-welfare-maximizing distribution of rights, then the problem would be merely a technical one, a problem of working out how to design the law to take photography into account. But if, like Bernard Edelman, one believed that they were arbitrary, that there was no correct answer, then the question was an interesting historical one: what confluence of political, economic, and ideological forces shaped how property rights in photographs were defined?

Edelman's analysis was significant because, instead of the standard easy and cynical answers to this kind of question—for example, the capitalist class or a coalition of interest groups simply gets what it wants—he granted that, in a specific way, ideas do matter. The law has to be *meaningful* in order to work, and the players in the game who made the law of copyright invoked the image of a laboring, creative individual getting his or her just rewards; in other words, coming up with an answer to copyright in photographs involved invoking a particular idea of the self or, in Edelman's terms, a subject. That subject was both depicted in the law and in a certain way enacted by its enforcement. Edelman, borrowing from French poststructuralism, understood the subject/self, not as a thing that automatically inhered in a person, but rather as a cultural accomplishment, a contingent organization of language and social practices, along the lines of what

John Frow calls "the imaginary forms of selfhood."[19] (The argument is, not that selves do not exist, but that they are not their own explanation; when someone deeply feels, "this is who I truly am" or when they behave according to a certain definition of selfhood—"I am a citizen," say, or "I am a businessperson trying to make a profit"—it's not that those claims are untrue, but that they have their own cultural conditions and thus cannot be taken as fully self-explanatory.)

Applied to the phenomenon of copyright, this meant that a sense of ownership of one's writings or efforts, a sense of responsibility for one's creative compositions, had to be acquired historically and culturally. It was not something obvious to all people at all times, as Ayn Rand might have it, but rather, as legal historian David Saunders puts it, a habit acquired "much in the same way as late twentieth-century Westerners have acquired a reluctance and incapacity to spit in public."[20] Part of Edelman's unique contribution was to take the idea of cultural subject-construction beyond the realm of literary analysis into a realm where one could specifically see the intersection of power, culture, and the state, in the moment of creating capital.

Some years later, across the Atlantic in the American academy, a Critical Legal Studies-influenced law professor named Peter Jaszi became interested in similar intersections, resulting in a then-unusual collaboration between Jaszi and literary historian Martha Woodmansee. In the wake of their collaboration, a critical literature on copyright developed that discovered, with a kind of astonishment, the romantic notion of the author-genius buried away inside intellectual property law.[21]

Copyright, historically a response to the capacities of the printing press, is not about property in a physical thing such as an individual book. It is about a text, that is, a sequence of words or an organization of colors, shapes, and sounds—something that can be reproduced across multiple instances, across the multiple copies of a book, a photograph, a film. But for this to make sense as something that can be owned, copyright needs to be granted to something the law can recognize as *not* a copy, something that is *original*, both in the sense of unique and in the sense of having an identifiable origin. That thing which is granted a kind of property status, then, has to be something that was not itself copied, something that had a moment of creation-from-nowhere. Beginning with printed books themselves, judges and lawyers, faced with legal squabbles and dilemmas, tended to imagine that original thing as something that sprang from a moment of inspiration inside the head of a unique individual, a genius. This, it turns out, was not a figure so much like Locke's yeoman farmer cultivating land or Bentham's calculating, profit-maximizing shopkeeper, but, both historically and phenotypically, it was something more like Goethe and Wordsworth's inspired romantic artist—the model for the romantic form of selfhood.

As a result, inside legal cases concerned with decidedly unromantic topics such as computer databases and genetically altered cells from someone's spleen, one can find invocations of something that looks very much like the shopworn literary figure of the romantic, isolated artistic genius working away in a garret. The literary critics and cultural historians found this interesting because they had been busy deconstructing the very notion of authorship. The signature essay, from their point of view, was Foucault's "What Is an Author?" which famously concludes with the question "What matter who's speaking?"[22] (Jaszi's introduction of this notion to the legal academy was called "Who Cares Who Wrote Shakespeare?")[23] The question has double implications. On the one hand, the question casts the common concern with specifying authorship into doubt: why should it matter who Shakespeare was as a person? What does that tell us about his works, about why they matter? But, on the other hand, it also raises a question of how the idea of the author as a genius-creator operates in history and society, the question of what Foucault called the author-function.

This approach opened the door to a great deal of fruitful scholarship that married the concerns of cutting-edge humanists and cultural critics with those of legal scholars. Film scholar Jane Gaines, for example, published a book demonstrating how an analysis of intellectual property can illuminate an understanding of films.[24] Law professor James Boyle analyzed trends in copyright law using insights borrowed from Foucault and other continental scholars.[25] Boyle in particular called attention to how the creation-from-nowhere assumptions associated with the concept of authorial genius, what he called the "author-ideology," had the effect of obscuring the social conditions of creation, leading to questionable legal policies and obscuring various forms of collective cultural and intellectual production. Law professor and anthropologist Rosemary Coombe further elaborated on the problem, granting Boyle's point but also noting the "double-jointedness" of the idea of authorship in law, the way it can go in multiple, sometimes unpredictable directions; if "author-ideology" generally functions to shift power over culture creation towards, say, Disney or Time Warner, it also can sometimes support, say, Native American groups trying to protect their cultural heritage.[26]

While scholars were pursuing these interesting questions, however, the courts and legislatures of the United States, and to a large degree of the world, were pursuing a decidedly less skeptical line of reasoning regarding private property. Under the sway of neoliberal habits of thought, property relations were being extended ever more widely—to water, to highways, to genes, and, in the realm of intellectual property, to software patents, to business models, to the "look and feel" of software, to ever-longer copyright duration—and in the early 1990s this was generally presented as the only logical approach. A task force created by the Clinton administration in 1993 released a White Paper calling for strengthening intellectual property in the face of new digital technologies.[27] In 1994, the United States and European nations succeeded in making intellectual property law an element of the international system of trade in the TRIPS agreement, administered by the WTO. Property in the digital realm was clear, efficient, moral, and—as far as those in power were concerned—inevitable. The principle of "the more property protection the better" seemed inexorable.

On the other side of the intellectual fence, it did not help that Critical Legal Studies and its fellow travelers were vague when it came to solutions. Critical Legal Studies was generally thought to be a left-wing movement because it typically crossed swords with both legal moderates and conservatives inside law schools and because it seemed to make a case for relatively radical changes in legal interpretation. It was not, however, exactly the activist Left of the civil rights movement of the early 1960s, which, following Martin Luther King, Jr., couched its claims in terms of the ideal of rights; that earlier version of the Left proceeded largely by demanding that the United States live up to its own ideals, that it uphold, as King said, every citizen's right to life, liberty, and the pursuit of happiness. Critical Legal Studies, in contrast, seemed to be saying, not that law was failing to live up to its own standards, but that it *could not* live up to those standards. Yes, this also meant that the law could in theory be changed any number of ways, but by itself the theory did not offer any basis for deciding what those changes should be. If this was true, moreover, what had happened to the rule of law? Were we just collapsing back into the rule of men, where judges settle disputes based on their own personal political views? Had we ever even left it? What was the alternative?

So, even for its enthusiasts, there was something disheartening about the Critical Legal Studies position. Once one had established, at least to one's own satisfaction, the bankruptcy of the dominant ideas, what next? Perhaps a colorfully ironic take on legal ideals might be enough for those who relish the moment of iconoclasm for its own sake. But, as conservative law and economics theorists were wielding ever more influence on actual legal decision making, for those with important insight into the holes in the conservative positions, a list of much-discussed books and a few tenured law professors making elegant critiques seemed like small comfort. If you wanted to do more than just take apart other people's ideas in front of your colleagues, if you wanted to be part of some actual positive change, if you wanted to do something that actually made a difference, where were you supposed to turn?

In this context, the surprising appearance of the internet presented an opportunity.

In March 1994, *Wired* published an essay by John Perry Barlow, which, in a dismissive sweep characteristic of both, was subtitled, "Everything You Know about Intellectual Property Is Wrong." Here, suddenly, was something that looked like a Critical Legal Studies argument appearing in a hip popular magazine. The resemblance, to be sure, was mostly in the title. The article, about patents and copyright in the digital realm, asked,

> If our property can be infinitely reproduced and instantaneously distributed all
> over the planet without cost, without our knowledge, without its even leaving our
> possession . . . what will assure the continued creation and distribution of such
> work? . . . The accumulated canon of copyright and patent law was developed to
> convey forms and methods of expression entirely different from the vaporous cargo
> it is now being asked to carry. It is leaking as much from within as from with-
> out. . . . Intellectual property law cannot be patched, retrofitted, or expanded to
> contain digitized expression any more than real estate law might be revised to cover
> the allocation of broadcasting spectrum (which, in fact, rather resembles what is
> being attempted here). We will need to develop an entirely new set of methods as
> befits this entirely new set of circumstances. . . . The source of this conundrum is as
> simple as its solution is complex. Digital technology is detaching information from
> the physical plane, where property law of all sorts has always found definition. [28]

For law professors and other intellectuals skeptical of the project of property, this looked like a friendly voice from an unexpected place. To be sure, for someone steeped in the critical and historical literature on intellectual property, many of Barlow's specific claims were dubious. Digital technology was different, but once upon a time so was photography, and Edelman had shown how successfully the law had been adopted to that once-new and befuddling technology; historians could tell similar stories about broadcasting, film, and recorded music. [29] And Barlow's idea that old laws were all based on "the physical plane" was peculiar; intellectual property has always been about *intangible* things. There was little new about the "virtuality" of digital property. (While it's true that copyright has generally protected only things that "are fixed in a tangible medium of expression"— that is, written down or somehow recorded—the difference between a printed book or a cassette tape and a digitally transmitted text is one of degree, not of kind.) Neither words displayed on a page nor words displayed on a screen are completely lacking in physical substance, and both are easily copied. The lack of fit between the law and the reality, from the point of view of the accumulated critical literature on property, was nothing new. A Critical Legal Studies aficionado might have said that everything we know about property *in general* was wrong, so what's the big deal about pointing this out in the digital realm?

Yet even for someone who was aware of all that, there was something alluring in Barlow's jeremiad. In a by-then-familiar move, Barlow distinguished between the cool young folks who got it and the old suits that did not. He argued that "most of the people who actually create soft property—the programmers, hackers, and Net surfers—already know this. Unfortunately, neither the companies they work for nor the lawyers these companies hire have enough direct experience with nonmaterial goods to understand why they are so problematic." The "programmers, hackers, and Net surfers," who at that time were already developing a heroically rebellious status in the culture, thus might in fact be willing to rally behind someone who claimed that all those old lawyers were wrong. Deconstructing property law might impress a few colleagues, but following Barlow's lead might take one to new, more hopeful places. Here was a potential ally, a new potential audience, and, unlike the seemingly impotent Critical Legal Studies, a hint of something that might actually *make a difference*.

Barlow's essay exemplifies how the generalized romanticized construction of the digital in the context of the spreading internet created a context for a popularized iconoclasm. The surprising spread of the internet with the romanticized sense of rebellion so successfully propagated by *Wired* suddenly made it easier, even attractive, to dismiss large chunks of the received wisdom. In this new context, intellectual property might be one of those areas where the received wisdom might be changed, not just criticized.

The Birth of Cyberlaw

Lawrence Lessig was at the time a promising assistant professor of law at the University of Chicago. He had published a series of articles on problems of Constitutional interpretation that demonstrated a solid awareness of the gnawing dilemmas of legal interpretation foregrounded by Critical Legal Studies, but an unwillingness to accept the fatalism the position suggested. His first law journal publication, for example, was a comparison and contrast between the theories of legal moderate Bruce Ackerman and those of the thundering neo-Hegelian star of the Critical Legal Studies movement, Roberto Unger.[30] Lessig argued that the two approaches were less distant from one another than was commonly assumed, which put him in the position of both young turk—each scholar and their followers were wrong about something—and moderate—the divisions were not as pronounced as most thought, and in the polarized legal academy of the day being a moderate could actually appear as iconoclastic.[31] And, as the 1990s progressed, Lessig published a series of scholarly articles that showed a keen interest in and grasp of the role of first principles and underlying assumptions in legal debate, while also maintaining a conviction that the law could be made to make sense, to live up to its promises; the law could work.

In the manner of Critical Legal Studies, Lessig regularly pointed to the weaknesses of the assumptions underlying various philosophies of jurisprudence, though he did so in a gentle tone. He wrote, for example, that the underlying principles of the entire law and economics movement tend to rely on an economic theory whose "sparseness and simplicity" sometimes might "make one miss something important."[32] Or that all the various efforts to define a system of Constitutional interpretation that accounts for the dramatic history of change in interpretations of the meaning of the Constitution rely on "plenty of intuitions, but no satisfactory account."[33] Unlike Critical Legal Studies, however, Lessig distances himself from those who would say that these weaknesses might make one wonder if legal reason itself is rotten to the core. An account of the meaning of human actions, he suggests, can complement efforts to apply economic reasoning to the law; a theory of translation can revive the idea of fidelity to the Constitution in the face of all the historical evidence that suggests a lack of faithfulness.[34]

But, rather uniquely for someone with his training, Lessig was also a habitual fiddler with computers, having dabbled in programming after college.[35] This put him in that category of individuals who would be experiencing the rise of the internet before their superiors caught on. While clerking for Supreme Court Justice Antonin Scalia in the 1990–1991 term, he managed to convince several justices to start using microcomputer-based publishing software to replace their antiquated system by demonstrating alternative software on his laptop. By Lessig's account, his Barlow-email-moment was a cover story in the *Village Voice* by Julian Dibbell called "A Rape in Cyberspace," which appeared in Dec. 1993—a few months after the internet first hit the media radar. It was a discussion of a virtual rape in an online game world.[36] According to Steven Levy, "as he read Dibbell's piece, Lessig was struck by how closely the concerns of the participants in the virtual world . . . resonated with those of [legal scholar Catherine] MacKinnon, whose radical views (porn isn't protected speech) were generally considered anathema at the *Voice*. This suggested to Lessig that cyberspace was virgin intellectual territory, where ideas had yet to be boxed in by orthodoxy. 'It was a place where nobody knows their politics,' says Lessig."[37]

The move from the dry world of theories of constitutional interpretation to writing about the internet, then, while certainly occasioned in part by Lessig's personal hobby of working with computers and the surprising explosion of the internet on the scene in 1993 and 1994, was also motivated by the desire to *do something that mattered* in a world where conservative thinking seemed to hold all the power. Looking back on his move towards becoming a "cyberlawyer," Lessig has said,

> There are issues I think are deeply unjust about our legal system, outrageously so. You know, the legal system for the poor is outrageous, and I'm wildly opposed to the death penalty. There are a million things like that—you can't do anything about them. I could go be a politician, but I just could never do something like that. But [cyberspace] was an area where, the more I understood it, the more I felt there was a right answer. The law does give a right answer.[38]

Lessig was not alone. In the early 1990s, among intellectuals spending their spare time discovering the pleasures of online communicating, the unexpected spread of the internet seemed to create an opportunity for bringing up big philosophical issues in fresh ways, in ways that might be heard outside the confines of a narrow circle of colleagues, in ways that actually might have an influence. The way that the internet took established institutions by surprise in the early 1990s offered an opening, a place where intellectual iconoclasm actually might gain some purchase outside the academy.

As we saw in the last chapter, libertarians like Esther Dyson began to discover this possibility in the late 1980s and seized on it in the pages of *Wired* and other venues; in a sense, their hope was that, somehow, computer technology could turn mud back into crystals. As the 1990s progressed, however, it was the iconoclasts of the legal Left who began to move towards the internet. Boyle, whose first book on intellectual property covered a full range of topics from indigenous cultural knowledge to genetics to insider trading, began focusing more on the specifically digital world in the mid-1990s. A legal historian on the faculty at Columbia Law School named Eben Moglen signed on to be general counsel to the then-little known Free Software Foundation in 1993; Moglen had worked as a programmer in the early 1980s, but his early legal career was made up of law journal articles on the historiography of early twentieth-century law (and an article that weighed in on some fine points of Critical Legal Studies's principle of legal indeterminacy).[39]

It is common enough for a mid-career professor, once granted tenure and thus no longer so needful of having to prove oneself to senior colleagues, to look for something a little more worldly, something that might take one closer to the rough-and-tumble of current events. But this move usually takes the form of backing off from the more abstract, perhaps philosophical concerns that seem of highest interest inside the academy; one starts accommodating the concerns of, say, politicians, or practicing lawyers, or interest groups. What is striking about the development of cyberlaw was the degree to which it was driven by a sense, not that one would have to abandon the philosophical to deal with the "real" world, but almost the reverse. The way the internet entered American social and political life created a context that seemed to actually welcome an inquiry into first principles while also maintaining a sense of positive possibilities. Intellectual radicalism—in the sense of a critique of the roots, of the underlying conditions of a situation—seemed to be the way to go. Maybe everything we thought about copyright (or property, or government regulation) was wrong; but, uniquely in the context of the internet, that conclusion was perhaps not dispiriting. It carried with it a sense that something could be done.

The Microsoft Problem

The first years of what would become known as cyberlaw mostly involved riding along with the *Wired* vision—the internet was a realm of potential new freedom that ought not to be sullied by old corporate and government ways—while debating the particulars with those earlier iconoclastic entrants into the open cultural space created by the internet surprise: the market purists and libertarians. However, as Microsoft readjusted itself to the emerging internet in 1995 and 1996, it no longer looked self-evident that the internet's openness would overthrow the Microsoft empire; Microsoft, by throwing its enormous resources into its browser and leveraging its domination of operating systems, could colonize the internet just as it had come to dominate the personal computer.

Hating Microsoft has been a popular sport, particularly among programmers and technophiles. The railing against Microsoft does not come from exactly the same place as the more generalized railing against corporations in the United States. True, ever since they took center stage in the U.S. economy in the Progressive Era, large corporations have been criticized and attacked almost as much as they have been celebrated and emulated. And intellectual property has often been a source of power for many of them; carefully cultivated patent libraries tied to investment-guided research and development provided much of the strength of early twentieth-century corporations, such as Dow Chemical, General Electric, Westinghouse, and RCA.[40]

But the resentment against Microsoft is uniquely personalized. Criticisms of Microsoft rarely associate the company with corporations with similarly privileged positions in their fields (such as Intel in microprocessor manufacture or Oracle in database management systems), and the structural conditions of corporate capitalism that render bigness its own reward are not usually discussed. Instead, criticisms are typically focused on the company founder, Bill Gates, and linked to criticisms of the quality of its products and the implication that there is something peculiarly nefarious about the whole enterprise. The website Slashdot, popular among computer professionals and open source enthusiasts, labels stories about most corporations—AOL, Google, Sony, Apple, Yahoo—with corporate logos or pictures of products. The icon for Microsoft-related stories, in contrast, superimposes the headgear of the Borg—the evil, mind-controlling empire from the *Startrek* TV series—on top of a picture of Bill Gates. Computer scientists who go to work for Microsoft find themselves apologizing for their choice of employer in professional fora. Like a mythic demon in a tribe's collective lore, Bill Gates has been turned into an important negative image in the symbolic economy of online culture.

Why this intense focus on the person of Bill Gates? Significantly, Bill Gates became the richest person in the world, not from software that he actually wrote, but basically by managing the labor of other software writers and carefully manipulating the distribution and marketing of their work. Gates did do substantial amounts of programming in the company's early entrepreneurial years of the late 1970s and early 1980s. But the products that became the foundation of the company's spectacular rise to dominance—MSDOS, Windows, Excel, Word—were created and maintained by others.

In the annals of corporate America, Microsoft's business strategy is hardly surprising. Silicon Valley journalist Robert Cringely claims that Gates once told him that "the way to make money in the computer business is by setting de facto standards."[41] Insofar as computing is about communication—which, as we've seen, has been mostly the case since about 1970—having the same system that everyone else has is its own reward. A better system that is different from everyone else's is, in an important sense, *not* better. Call it network externalities, call it path dependence, or call it the fundamentally social character of human existence, there is much value in commonality, and Microsoft's core strategy has been to exploit that fact. Become the norm and then charge for it; all other goals are secondary. MSDOS was probably not technically better than other available operating systems of the early 1980s, but Microsoft did everything they could to insure that it was everywhere. And as it spread everywhere, a cycle was established that encouraged manufacturers of hardware and software to create things that worked with Microsoft's operating system, and as they did so they further reinforced Microsoft's position; this self-reinforcing cycle continues to this day.

From a business management point of view, this is not all that troubling. *The Economist* calmly noted that, in spite some of his claims to the contrary, Gates is not a technological leader who sees further ahead than others or builds unique and pathbreaking products. His skill is in "making money in the slipstream of other people's technological vision"—and this on its own is perfectly reasonable, as far as *The Economist* is concerned.[42]

Bill Gates is uniquely a problem only if you think that the acknowledgment and reward of individual effort and creativity is of overriding importance —that is, if you are attracted to a romantic individualist point of view. From this point of view, Microsoft is incredibly galling, particularly to someone who actually does do the work of managing or making software. As Microsoft's dominance steadily grew in the 1990s, the growing world community of programmers and other computer experts regularly experienced visceral annoyance at the success of Bill Gates; he was reaping the rewards, turning those rewards to his further advantage, but he wasn't doing it by building a better mousetrap, and he wasn't doing what they saw as the real work. He wasn't creating much that was truly new or better. If the most fundamental of all rights is, as Ayn Rand put it, "to think, to work and keep the results," there was something wrong; the people who were doing the thinking and working weren't keeping the results. This situation was not enchanting.

The Creation of the Open Source Idea

One of the difficulties for Bentham's premise that people are rational and self-interested is that, on the surface of it, they sometimes are not. Most people at times clearly like to do work for something larger than themselves, for example. Soldiers, athletes on teams, mountain climbers, family members in moments of celebration or crisis, and law students on the law review often recall the intense energy and feeling of solidarity that comes from working on behalf of the group, frequently with an amount of effort that is clearly not in one's obvious self-interest. People will often say, in fact, they have worked harder in these situations than in contexts where they were working purely for their own gain. It is only because utilitarian reasoning has become so taken for granted in our culture that the same people who speak fondly of such moments of working for the group can turn around and say things like "everybody knows that welfare discourages initiative" or "if people can't make money from it, it won't get done."

Yet more than a few computer engineers know from personal experience that sometimes people will do things even if they could make more money from doing something else, and, as we saw in chapter four, a number of those inside the internet engineering community saw computer networking as a case in point. There are times when some of the best work is done, not to maximize profit, but out of passion or commitment to something larger. In the early 1990s, as the internet spread like wildfire and became coupled in the popular imagination to an unregulated free market, some of these engineers started to raise their voices in protest. Michael and Ronda Hauben, for example, published *Netizens*, an important book that compellingly detailed the numerous ways in which the internet was born of and embodied, not capitalist self-interest, but forms of spirited and deliberately collective action. Unix, they showed, was designed from the ground up to enable collaboration and the sharing of interoperable software tools so as to encourage collective improvement of the system. The internet appeared, not just because of various nonprofit arrangements, but because of its deliberate design, on both a technical and social level, as an open system built for shared collective effort. Usenet and other nonprofit collaborative communication systems both spread much of the knowledge that made the global internet possible and taught a generation of technical professionals the value of online, citizenly collaboration.

It is perhaps because of the overwhelming dominance of the taken-for-granted utilitarianism in American society that those with different views sometimes feel compelled to leap to the other extreme. The Haubens took their crucial observation—that some of our most advanced technologies like the internet did *not* emerge solely from self-interested, profit-motivated contexts and logics—and used it to then insist that nonprofit behavior is obviously morally and technically superior. They defined and valorized the "netizens," not as people who merely used the internet, but people who "understand the value of communal work and the collective aspects of public communication ... people who care about Usenet and the bigger Net and work towards building the cooperative and collective nature which benefits the larger world."[43] "The so-called 'free market,'" they argued, "is not a correct solution for the problem of spreading network access to all."[44] Never mind the ideologically blurry tradition of Vannevar Bush, with its technocratic faith in the ability of experts to elegantly mix public and private efforts or all the back-and-forth movement between private and public contexts characteristic of many of the engineers that the Haubens lionized. In the first half of the 1990s, faced with the spread of the bizarre claim that the triumph of the internet was somehow a triumph of free individual initiative and of the market, the Haubens seized on the opposite claim: *Netizens* presents a picture of computer communications as a nonprofit communitarian utopia, threatened by ignorant capitalist managers.

This stance of reacting to utilitarian dominance with communitarian purism is shared by Richard Stallman. Today he is a hacker hero, but in the early 1990s his effort to make a free and open clone of the Unix operating system was known only to narrow computer engineering circles (and those readers of Levy's *Hackers* that made it all the way to the end). Stallman, upset by the restrictions that ensued from the commercialization of systems like Unix in the late 1970s and 1980s, set out to create, not just another version of Unix, but an alternative software universe in which the free sharing of software code was *required* by those using it; towards that end he created the Free Software Foundation and a unique form of copyright license, the General Public License or GPL, that specified that a piece of software could be freely shared and used provided that whoever distributes it also freely distributes the source code and any modifications to that source code. Instead of, say, releasing software code into the public domain, copyright was retained, but for the purpose of maintaining free and open distribution, not for the purpose of preventing others from selling the software. It was less a nonproperty or public domain approach than a kind of antiproperty. In 1991, when a Finnish college student named Linus Torvalds, as a way to learn about operating systems, began building an experimental operating system kernel, he early on copyrighted it using the GPL, so that he and collaborators could also use software tools created by Stallman and his colleagues.[45]

In research and academic environments, where one's prestige and job security often depends on open publication and collaboration, this made a certain kind of sense. But in a neoliberal world that was both in love with high technology and that seemed completely stuck on the assumption that innovation only sprang from the unfettered pursuit of profit, Stallman's approach was so different as to be almost invisible. In its second issue in 1993, *Wired* did run an article titled "Is Stallman Stalled?" that briefly summarized Stallman's approach. The author quotes Stallman, "I don't think that people should ever make promises not to share with their neighbor," and then spends close to half of the article discussing problems with Stallman's effort. "Things seem to have bogged down," the article states, noting delays in software production and problems with financing. The tone, particularly in the context of *Wired* at the time, is that of the knowing adult speaking of an idealistic child. If there is any success here, the article suggests, it is in the for-profit efforts to support Stallman's programs, like Cygnus Support. The article is subtitled, "One of the Greatest Programmers Alive Saw a Future Where All Software Was Free. Then Reality Set In." "Reality," the reader must assume, is the world of self-interest, profit, and the market; sharing is a naive ideal. *Wired* was in effect telling its readers to move on, nothing to see here; the action is all in the romantic entrepreneurialism of start-ups.

In the next few years, as the business community took *Wired*'s admonition to heart and plunged the American economy into the stock bubble, those with doubts about the "new economy" mantra quietly went about their business, while looking on helplessly at the dotcom rush. Stallman, Torvalds, and many other engineers continued tinkering with their systems. Scholars like Lessig and Boyle kept looking for ways to use the new technological situation to justify rethinking some core principles about law and society. The Haubens and others enamored of the communitarian moments at the internet's roots kept promoting their vision, most often on the internet itself. But the world still seemed completely enamored of markets. The stock market continued to rise, and market practices proliferated wildly across the world, appearing successfully in places never before thought possible like Russia and China and in industries like telecommunications, where the market-driven mobile phone rapidly leap-frogged the regulated or state-owned land line telephone systems in many parts of the world. The many and crucial nonproprietary aspects of the internet, therefore, remained largely ignored. In spite of the best efforts of those like the Haubens, the internet's history of nonprofit development, the many aspects of its development that involved various degrees of deliberate openness, the fact that much of its rapid success had to do with its open, welcoming structure and design: all these details could not penetrate the fog of the market enthusiasms of the early 1990s.

But there remained the problem of Microsoft. Microsoft's dominance was still galling from the point of view of the programmer who wanted to believe that programming was something like art deserving of recognition. The stock bubble, as we have seen, could not have existed without the narrative of the romantic programmer; it would not have been compelling enough to take off without all the tales of rebel-hero programmers like Andreessen, which offered the hope of fusing the pursuit of wealth with the pursuit of self-expression and rebellion. Yet, as the 1990s progressed, Microsoft only increased its control over the operating system market and by 1996 started making dramatic inroads into Netscape's share of the internet browser market. Major corporations like IBM and Apple found themselves gasping for breath in the face of Microsoft's dominance.

What it took to sever the dominant culture's association between the internet and the neoliberal market was Linux. By 1994, Torvalds and fellow tinkerers improved their new operating system kernel to the point where it could be combined with Richard Stallman's work to make a rather effective complete software system, which began to look attractive to some software professionals. On the one hand, it was a familiar system to many experienced engineers already working in the area, as it was derived from Unix. On the other, it overcame some of the problems associated with the existing fragmentation of Unix into competing systems, and, because it was open, it allowed for easy tinkering and improvement. Finally, Linux went through a period of rapid technical improvements at the same time that Microsoft's operating system dominance became so complete that it could afford to release software that was merely good enough in areas where it dominated and focus its vast resources on areas where it had competition, like the browser market. The average consumer booting up Windows 95 would not likely notice. Computer engineers, however, did; beginning with sheer technical admiration for Linux coupled to annoyance with the technical limitations of Microsoft's software, they then began to wonder what the moral of this story was for the economic organization of software production.

This was the context in which a Unix programmer and gadfly named Eric S. Raymond wrote an essay called "The Cathedral and the Bazaar," which he delivered at programming conferences in 1997 and which went on to circulate beyond the internet into the offices of key business executives and copyright lawyers, initiating a ground shift in corporate strategy. This article was passed around at Netscape in the run-up to its decision to open source its browser in January 1998. Not long after that, Apple would decide to base its next generation operating system on an open source foundation, and IBM and Sun Microsystems would adopt open sourcing as a key strategy.[46] Soon, Linux-based companies like Red Hat would become stock market darlings.

It is significant that the arguments of "The Cathedral and the Bazaar" are not communitarian. In contrast with Stallman or the Haubens, Raymond dismisses appeals to altruism out of hand.[47] The central rhetorical accomplishment of the piece rather is to frame voluntary labor in the language of the market; the core trope is to portray Linux-style software development like a bazaar—the archetype of a competitive marketplace—whereas more centralized and controlled software production is portrayed as hierarchical and centralized—and thus inefficient—like a cathedral: static, inefficient, medieval. (While Raymond seems to have had previous efforts in the Unix world in mind when describing cathedral-style software development, it is a safe bet that many readers thought of Microsoft.) Raymond thus disarticulated the metaphor of the market from conventional capitalist modes of production and reconnected it with a form of voluntary labor, of labor done for its own sake.

Part of what makes this curious reversal work is Raymond's focus on programmers' motivations. For an essay about such a dry and technical topic as the management of software development, there's an awful lot of reference to the internal feelings, psychological makeup, and desires of programmers. (Subsequent discussion of open software seems to have maintained some of this focus.)[48] Almost like a hip *Entwicklungsroman*, Raymond presents a first-person account of his own experiences in software development, during which he tells the story of how he became converted to the Linux software model. This narrative of personal revelation is interspersed with numbered principles or aphorisms, the first of which is: "*Every good work of software starts by scratching a developer's personal itch.*"[49] Because Raymond's audience is in the worlds of business and law, he immediately sets out to reconcile his psychologically tainted portrayal of motivation with a utilitarian one. "The 'utility function' Linux hackers are maximizing," Raymond continues, "is not classically economic, but is the intangible of their own ego satisfaction and reputation among other hackers." Raymond goes on to draw analogies with fan subcultures, wherein the enhancement of reputation among the other members of the community is understood as a key motivation.[50]

Much of the piece is devoted to explaining why a wide-open approach to software development involving frequent borrowing and sharing of code, early and frequent releases of updates that have the effect of involving users in development, and attention to maintaining positive social relations among participants all combined in the case of Linux to create better software. But it's crucial to the essay's effect that Raymond frames the motivation to write software as something born of a not entirely rational fascination or ambition (an itch), of a desire to have one's accomplishments recognized not with money but with the psychological satisfactions of acclaim. One could of course criticize this as both an empirical description and as a philosophical argument, but what's significant is, first, how the dream of having one's "itch," one's inner passions, acknowledged by a community of the like-minded is a characteristically romantic structure of feeling and, second, how much Raymond's statement of the problem, whether or not it is philosophically coherent, resonated with the computer culture at large and had some impact on the larger business culture in a way that communitarian or managerial arguments have not. In the minds of quite a variety of people, this vision of passionate programmers provides a much more appealing way to deal with the perennial industrial problem of monopoly than something like industrial policy or antitrust law.

Raymond publicly presented "The Cathedral and the Bazaar" in September 1997, after which the essay began to circulate. At the time, the executives at the Netscape corporation, after riding high on the early stock bubble, had been sensing doom. In 1997, they had been watching Microsoft's share of the web browser market rise from negligible to almost half of the market, while Netscape was making almost no income and Microsoft's profits from its other software was setting records.[51] So in January 1998, they announced they would publish the source code for the browser for free distribution, and in February they invited Raymond and several others to a meeting to help them plan their new strategy. Those in attendance at the meeting saw this as an important opportunity to get the corporate community to take the free software community seriously and towards that end chose to follow a pragmatic path of using the term *open source software* (instead of Stallman's term *free*) and of emphasizing technical advantages rather than ideals—which qualifies as an act of rhetorical genius. A few weeks later, Raymond and others would found the Open Source Initiative to support these efforts, and the organization would be involved in strategy shifts at Apple, IBM, and other companies.[52] Today, Linux and other forms of open source software are central to many different businesses world wide and can be found on everything from cell phones to massive research computers. If one had predicted this chain of events in 1994, one would have been dismissed from almost all directions as hopelessly naive.

The argument here is not that this single essay by itself directly caused major corporations to adopt new strategies; rather, Raymond's essay helped promulgate a way of understanding software development that played a key role in the corporate shift. It was a necessary but not sufficient part of the conditions of possibility of the move towards open software. Of course, these companies had an economic interest in the new strategy, especially given the Microsoft monopoly. But the economic conditions behind the change had been in existence for several years; an economic explanation alone cannot explain why these companies made their policy changes all within a roughly one-year period (1998).

In different times, open source might have seemed unremarkable; when Bell Labs gave up its patents on the transistor as a result of a consent decree in the

1950s, this was viewed as a reasonable solution to a complex problem, not egregious theft. But in the first half of the 1990s, paying for labor that produced something that could only be given away for free would have been considered deeply irrational. In 1997, the embrace of open source—as modest as it was—marked a profound shift in the common sense of U.S. political economic thinking in the high-tech realm. After the arrival of the open source movement, the neoliberal assumption that more-property-protection-is-better was no longer unassailable.

So how did this happen? Traditional economic reasoning does offer explanations of why a company might use open source software. A consumer device manufacturer like Tivo makes its money selling devices, not software, so using Linux makes sense not only because the software is free but because its open character allows Tivo to easily modify it according to its needs and to draw on the global support of the Linux community. Companies like Red Hat or IBM can make a solid profit by offering technical support to companies using their software, while allowing the software itself to be freely copied and shared. And when faced, as many high-tech companies in the late 1990s did, with the huge market power of a company like Microsoft, taking a gamble on open source software might have seemed like the only alternative. The only way to compete with Microsoft's market power was to offer a platform that was free to consumers and that, because of its free and open character, could create a united front among Microsoft's competitors.

But these economic reasons for turning to open source cannot explain why the shift happened when it did. All of these economic factors were in place in 1994, and yet no major company even briefly flirted with the idea of open software at the time. It was not until the fall of 1997, with the circulation of Raymond's essay, that the idea could even begin to get attention, and then in 1998 and 1999 the concept all of a sudden became relatively mainstream. It took Raymond's articulation of free software within a romantic individualist structure of feeling—and its appearance against the backdrop of the Microsoft problem—to lay the conditions for mainstream acceptance of the idea.

The acceptance was not instantaneous, of course, and as of this writing is not universal. Libertarianism of whatever variety is premised on the idea of private property, and so it is not surprising that many of the libertarian faith at first scoffed at open source. For example, in late 1998, once open source had gained some attention in the media, Wayne Crews of the Competitive Enterprise Institute published a critique of the open source movement (as part of a series titled "C:\spin: An Occasional Commentary on Regulation of High Technology—From an Undiluted Free Market Perspective"). Arguing that "like free love, open-source code is fun, but it's probably not a way to run the world," Crews wrote that, for the most part, the prospect of becoming fabulously wealthy, not the desire to give things away, drives software innovation. Nearly all "freeware" programs—whether word processors, image editors, games, or browsers—pale beside superior commercial versions. Even Netscape's release of the source code for Navigator—applauded by the open-source advocates—wasn't a fundamental embrace of their doctrine, but an effort to create a pipeline for offering other, more profitable services. Conveniently ignored also was that the Netscape giveaway occurred after the IPO that made multimillionaires out of its founders.[53]

Other critics were even less circumspect. Bill Gates and Steve Ballmer at different times insinuated that there is something communist about Linux.[54] *Forbes* magazine scoffed at the open software "movement's usual public image of happy software proles linking arms and singing the 'Internationale' while freely sharing the fruits of their code-writing labor."[55]

But such dismissals came across as shrill and tinged with desperation. They did not have the same kind of force that they might have had earlier in the same decade. Here was a technically sophisticated operating system that in some contexts seemed better than Microsoft's products. As Windows users grew accustomed to the "blue screen of death"—the end result of a system crash in Windows, a frequent occurrence in the mid-1990s—the sheer technical quality of Linux stood as a glaring refutation of one of the central claims of the neoliberal argument about intellectual property; here was better software created *without* the incentives of property protection. And, after elevating disheveled programmers to the status of cultural heroes in the earlier 1990s—remember that it was Netscape's publicity strategy to foreground its young programmers as opposed to its managers—the culture found it was exactly those heroes who were now increasingly celebrating something that seemed to point in a very different direction. In the intellectual space created by readers and writers of *Wired*, one had to recognize in Raymond's rendition of open source much of the same Byronic attraction that had driven the magazine's rhetoric in its earliest days; scoff at open source and you might start to look like one of John Perry Barlow's dinosaurs, one of the old suits who didn't get it. The romance of open source was too alluring to ignore, and the previous few years had taught the culture that this might not be a fleeting trend; the romantic allure of the hacker had already influenced the global economy by way of its role in the stock bubble.

The full intellectual structure of a point of view is often revealed, not just in the statement of high ideals, but in what people define as pragmatic—in those moments when someone claims that it is time to be sensible, to make compromises, to take a middle road. So, for example, in response to Wayne Crew's market-based dismissal of open source, fellow libertarian Esther Dyson offered a path towards reconciling with the movement. "There's a fundamental misunderstanding here," she wrote,

Open source software may be freely available, but someone *is* responsible for it. Most of the support for OS software is paid for; that's how (many of) the hackers' paychecks are funded. There is a lot of value—and money—floating around the world of OS. And yes, Netscape's use of OS to make its other services attractive is a legitimate, acknowledged and sensible business model.... (It seems to me that there are religious extremists on both sides of what ought to be an argument about business models, not morality.)[56]

If two years earlier Dyson felt comfortable stating that the market simply "works and is moral," now, in her mind, it was time to move away from blanket statements about morality. Now it was about business models, about being sensible, about striking a middle path. This tone of moderation would allow someone of Dyson's convictions to maintain a friendly public stance towards open source.

But that tone of moderation would also allow the loosening of the link between the romantic view of individual freedom and market libertarianism. For the first time in more than a decade, it became possible in business culture to seize the glamorous position of rebellious high tech while also supposing that, Thatcher notwithstanding, perhaps there *is* an alternative to the market.

Slashdot and "Code Is Law"

In the late 1990s, the website Slashdot did for open source what *Wired* magazine did for computer entrepreneurialism in the early 1990s. Created in 1997 by Linux enthusiast, student, and part-time commercial website developer Rob Malda, Slashdot evolved from a small website for listing and discussing technical issues mixed with personal anecdotes into the leading forum for open source enthusiasts, playing no small role in establishing a cultural tone for the movement and helping to communicate that tone to the rest of the constantly expanding web-surfing world. Slashdot's title banner describes it as "News for Nerds. Stuff That Matters" (as if in mockery of the self-confident grandiosity of the *New York Times* subhead "All the News That's Fit to Print"). Created before *blogging* entered the lexicon, Slashdot presented itself as self-consciously eccentric and as an entry-point into a labyrinthine world; updated around the clock, each front-page "story" on Slashdot consists of a short paragraph of "news" followed by an always steadily growing train of posts from reader/contributors. Scrolling is a necessary part of the experience, as is following links; the familiar slashdot-effect refers to the highly predictable overloading of external web sites within minutes after their URLs are posted in Slashdot stories.[57] While much of Slashdot's content concerns open software fairly directly—new Linux software releases, developments in intellectual property law—a good deal of the content is more general, reflecting the interests and spirit of its young coding-adept or technologically fascinated producers and readers: intriguing developments in science, reviews of science-fiction films, amazing things done with Lego.

Like *Wired* several years before, much of the thrill of Slashdot comes from the implication that you, the clever Slashdot reader, stand apart from despised others in the world—from the drones in suits who work for Bill Gates, for example. The implication, then, is that by reading Slashdot, you are part of a distinct cadre; the community is very much defined in terms of its opponents. While enamored of open source, the ethos is not particularly communitarian or somberly political. It is perhaps not accidental that the term *slashdot* derives from the keyboard command "/." that takes the operator to the root directory of Unix systems, a privilege only available to system operators with absolute superuser privileges over a multiuser system. It's a common command if you're fiddling with the technical setup of a Unix system. But it's also about power. If you can type "/." on a Unix computer and get to root, you can get into and modify anyone's account on the system. You can do things like erase the entire hard disk, or read other people's email, or change their passwords. You are, in the narrow world of that computer, omnipotent. Slashdot feels more like a band of misfit heroes than a Quaker meeting.

While the open software movement had been quietly gestating, legal intellectuals like Lessig and James Boyle had been exploring the intersection of intellectual property concerns with the internet as a way to bring fundamental questions about the law into broader recognition. Boyle, for example, moved from law journal articles about theories of legal interpretation to a series of pieces about internet issues such as privacy, censorship, and intellectual property. Well schooled in the debates about the ambiguities and limits of the category of authorship, Boyle tended towards an emphasis on the limits and blind spots of a strictly individualist, rights-based approach to law and technology. Early in his career he published an essay on the limits of the idea of individual subjectivity, bringing Foucault's critique of the subject into critical legal theory by showing how the idea of subjectivity itself is an unstable category, an effect rather than cause.[58] In his 1996 *Shamans, Software, and Spleens* he described the use of the romantic author construct in intellectual property law as an "author ideology" that blinded its adherents to the often collective sources of cultural innovation. In 1997, he published an essay criticizing the underlying assumptions of *Wired*-style digital libertarianism; calling for an internet analogue to the environmental movement that focuses on the structures and strictures of emerging intellectual property regimes, he has sought in various ways to emphasize the importance of actively supporting the public domain and shared culture more broadly.[59] In general, he pointed towards kinds of civic republicanism as an antidote to what he portrays as the blinkered and short-sighted radical rights-based individualism that motivated both trends in intellectual property law and much of the thinking about the internet in the 1990s. Maybe there is more to life than Emerson's autonomous self after all.

Larry Lessig started out on a similar trajectory, moving from theories of legal interpretation into the worlds of internet law and intellectual property, similarly bringing with him a sharp sense of the way that legal rights can be as constraining as they can be freeing. Lessig's best known book, *Code and Other Laws of Cyberspace*, is based on what looks like a classic legal realist maneuver; by pointing to the regulatory character of various private activities—in this case, coding—one undermines the common assumption that narrowly defines freedom as the opposite of government action. A simple demand that government keep its hands off the internet, Lessig patiently explained to his internet-fascinated readership, was no guarantee that the internet would remain free. In this Lessig was in keeping with the tradition of legal realism that also influenced Boyle, Jaszi, and others.

Unlike Boyle, however, Lessig did not pursue the Foucauldian critique of the subject and its interest in the limits and conditions to the idea of a free individual. For Lessig, the free individual was still very much the goal; his argument just pointed to the ways that private as well as government efforts could limit freedom. Open source software is "a check on arbitrary power. A structural guarantee of constitutionalized liberty, it functions as a type of separation of powers in the American constitutional tradition. It stands alongside substantive protections, like freedom of speech or of the press, but its stand is more fundamental."[60] If Boyle was calling for his readers to abandon an obsession with the abstract free individual and start thinking more complexly about the social conditions that support innovation and culture, Lessig presented the choice as a simple, stark one: Lessig titled one essay, "An Information Society: Free or Feudal?"[61] While standing alongside Boyle in attacking the libertarian notion that markets and private property are the sole guarantors of freedom, Lessig seemed to concede to the libertarians one thing that Boyle did not: the idea that freedom itself is a simple condition, an absence of constraint, the ability of individuals to do what they want, especially to express themselves, to engage their creativity. Boyle approached the romantic ideal of individualism skeptically. Lessig embraced it.

A search of the last decade of Slashdot postings for the name "James Boyle" turns up about fifty hits. Lessig turns up over one thousand.[62] Lessig's large readership is the product of many factors, not least of which are his immense talent, persistence, productivity, and character. But he also writes and speaks in ways that are carefully tuned to audiences like Slashdot, for whom the free individual is understood in romantic terms: as someone who creatively expresses themselves, often against the powers that be, and gets acknowledged for their accomplishments. Boyle is arguably wise in asking his readers to think beyond an obsessive focus on an abstract individual freedom. Lessig, however, by choosing as a starting point to emphasize the gap between a romantic individualism and the utilitarian one, has found a framework that resonates with a larger audience.

Conclusion

In a crisp essay, Milton Mueller has neatly debunked the grandiose claims occasionally made by both the supporters and opponents of open source software, that it is somehow a threat to capitalism in general. Mueller argues for a more pragmatic approach, which focuses on open source as a means to the end of individual freedom, not as an end in itself.[63] This is entirely reasonable. But the very fact that Mueller can be effectively making that argument today is in part due to the romanticism, with all its grandiosity, that became attached to open source in the second half of the 1990s and thus propelled the phenomenon into the limelight. As of this writing, the world of intellectual property law remains turbulent and contested. But this contestation marks a remarkable change from the legal and political atmosphere of the early 1990s in which intellectual property expansion was imagined in the halls of power as inevitable and self-evident, as not worth arguing about.

Since the rise of the open source movement, essays sympathetically expounding the ideas of someone like Lessig or Boyle have appeared in mainstream outlets like *The Economist*.[64] Business executives in many industries express an interest in a major rethinking of the patent system. Even the recording industry—once the leader in a hardline approach to copyright enforcement—is substantially softening its position. Some major record labels, for example, are now offering much of their content for download in a noncopy-protected MP3 format, an act that in 1995 would have been seen as childish folly. Open source software is now understood as a reasonable technical option in many contexts worldwide, and Linux continues to quietly spread, running on servers at web search firms like Yahoo, on cell phones made in China, on digital music players made in France, and on personal computers sold at Wal-Mart. In 1999, the original romantic copyright protectionist, Ted Nelson, open sourced the ongoing Xanadu project, "in celebration of the success and vast human benefit of the Open Source movement."[65]

The language of open source and its associated ideas, moreover, has been seized on in other domains. The use of the term *open* to refer to nonprofit decentralized efforts—the construction first seized on in 1997 by a handful of programmers as they groped for a terminology that would help legitimate nonproprietary software practices to business management—is now spreading throughout the polity. Not surprisingly, this trend began in technical areas. In 2001, with much fanfare, MIT reversed the 1990s trend in higher education of trying to commercialize educational materials and courses on the web by announcing what it called Open Courseware, an initiative to put all of its course materials online in a way that was free of cost and available to the worldwide public. A group advocating radical new approaches to radio spectrum management adopted the term *open spectrum.* (A White Paper describing the approach echoes Barlow in its first line: "Almost everything you think you know about spectrum is wrong.")[66] But the trend has expanded into areas where it is not just about technology. Advocates of decentralized, grassroots political action crow about the rise of "open-source politics" during Howard Dean's run for president in 2002 and 2003.[67] Critics of mainstream media advocate and explore "open source journalism" as a more democratic alternative to conventional journalism.[68] Brazil's minister of culture—a former dissident and popular musician—cites both Lessig and countryman Roberto Unger as influences in his "Culture Points initiative," which gives grants to local artists in poor areas to cultivate emergent local genres such as Brazilian rap music.[69]

Someone like Esther Dyson might argue that these trends are simply about another business model or that calling for the free distribution of information is hardly a new idea. She'd have a point. Universities and libraries have often in various ways supported the free and open distribution of information as a matter of organizational principle. And open source by itself is hardly a threat to capitalism as a whole. Any thorough look at the history of capitalism shows that "pure" markets have at best been temporary and fleeting events; capitalism has generally thrived only in the context of various extra-market political and institutional underpinnings, with some things treated as property amenable to exchange and other things not.[70] All economies, it turns out, are mixed. If, say, operating systems become all open source, if they are moved from the category of things that are exchanged into the category of noncommodified things that enable other things to be exchanged, capitalism will not come crashing to the ground.

But the role of open source as a political economic object lesson cannot be dismissed. Capitalism may not require pure markets or crystalline property relations, but it does need some kind of legitimacy, some mechanism by which it can be made to *feel* right, or at least worth acquiescing to, among broad swathes of the population. Romantic individualism, understood as a structure of feeling mapped onto a mix of experiences with computer use, is, as we have seen, a persistent phenomenon in American culture, one that has its own specific character and valences. If, in the early 1990s, *Wired's* version of romantic individualism was harnessed to neoliberal market enthusiasms, later in the decade that same structure of feeling, as articulated by Eric Raymond, Larry Lessig, and Slashdot, became a key element in a countervailing effort. At this point, the details of that object lesson remain confused and blurry. But the assumptions that dominated decision making regarding intellectual property in legal and managerial circles from 1980 to 1997 have changed; it is no longer automatically taken for granted that property protections are the best or only incentive for technological innovation, that stronger and broader property protections are always better, and that a digital economy could or should rest centrally on the commodification of information. Before 1997, critics of this common sense were not so much rebutted as ignored. After the rise of the open source movement backed by the intellectual work of the cyberscholars, they no longer could be, and that shift happened, in part, because of the widespread circulation of the romantic celebration of software creation as a form of personal expression.

And this may go beyond *intellectual* property. Property itself, as Carol Rose put it, traditionally has functioned as "the keystone right," in the American legal tradition, serving as the model for the very idea of liberty.[71] As a consequence, property rights have tended to trump all other rights, such as free speech rights; the rights of the owner of the shopping mall or the newspaper generally outweigh the rights of an individual speaker who is visiting the shopping mall or working for the newspaper. This pattern has been embedded in legal decision making in the United States for most of the twentieth century. Yet, in the last decade, the open source movement has occasioned a rethinking of that impulse by demonstrating in vivid ways how overly strict protection of property rights can conflict with the rights of speech and self-expression. In time, the open source movement may be the starting point for a significant loosening of the link between property and other forms of freedom in the American psyche.

Conclusion

Capitalism, Passions, Democracy

BACK IN 1994, towards the end of the stock-bubble-inspiring *Wired* interview with Marc Andreessen, interviewer Gary Wolf pressed Andreessen to specify the exact difference between what he was doing and the efforts of Microsoft, whom Andreessen saw as "the forces of darkness." Is not Netscape, Wolf asked, also a for-profit software company seeking to dominate a market by establishing a proprietary standard? (Netscape was giving away the program, still called Mosaic at the time, but not the source code, and it was rapidly creating proprietary new standards for web content.) When Wolf pressed Andreessen on these issues, after some waffling, Andreessen replied: "The overriding danger to an open standard is Microsoft . . . [but] one way or another . . . I think that Mosaic is going to be on every computer in the world." Wolf waited for more. Andreessen repeated himself, "One way or another."[1]

It is to Wolf's great credit that he ends the article there, with an ambiguity, leaving Andreessen's repeated "one way or another" hanging in the air. This is a telling moment. To be sure, it is not a utopian one. The young Marc Andreessen, full of bravado and brash ambition, can hardly be seen as a proponent of alternative modes of production. But it is not a case of false consciousness, either. What is significant about the passage is that *even* Andreessen, the man at the center of the stock bubble, speaking at the beginning of an historic moment of astonishing triumph and ideological hegemony for free market global capitalism, would feel compelled to say "one way or *another*," would so easily take for granted that there could be *another* way. This is another bit of evidence of the fact that the fabric of American cultural common sense, with its romantic threads, is open to alternatives to the market, even at moments when America's dominant ideologues are not.

Like Wolf, critical thinkers need to listen to the "or another" in the general discourse, to stay attuned to the tensions in various narratives of technological liberation, and treat them as part of the lived reality of the moment, not merely as something for which the only choice is celebration or denunciation. Wolf's obvious love for computer romantics should not be reduced to some kind of ideology or foolishness, and yet his sense of irony is worth pointing to and serves as a useful starting place for putting romanticism into a larger context.[2] That has been the approach of this book: to give romanticism respectful attention, though not uncritical acceptance. The significant thing in the first instance is that romanticism has both a persistence and effectiveness in U.S. culture that matters.

True, the hacker ethic does not add up to a coherent set of principles for organizing software production, and it was folly to believe those who claimed in the 1990s that the internet had changed the laws of economics. Yet there would be no President Obama without the internet,[3] the parameters of intellectual property in law and politics have changed dramatically since the beginnings of the open source movement in 1997, and the idea and practice of grassroots democracy have gained a new cogency in U.S. and global politics in association with the internet. The necessity of and problems with supporting private media with advertising have been more widely called into question than at any time since the 1960s. And the very idea of freedom has been distanced from property rights in new and significant ways, opening up possibilities in the United States that have not existed since the Progressive Era. No, this is not the revolution, and, yes, in the mid-1990s it was all harnessed to breathe new life into neoliberalism. Yet the internet has been at or near the center of some of the most significant set of ground shifts in American political practice since the Reagan revolution in 1980 and the dissident movements of the 1960s (which weren't the revolution either, but which did change the country in a variety of lasting ways).

This concluding chapter, then, explains how this could be so by summarizing the findings of the book and offers some observations about what the past decades can teach us about the relations between capitalism, technology, culture, and everyday life. It makes the case that the internet is open and disruptive, not because of anything inherent in the technology, but because historical circumstances allowed it to be narrated as open, because the stories that have become common ways of making sense of it have represented it as open, and in turn those stories have shaped the way it has been embraced and developed. The internet is potentially open because people have made it so, and there is a lesson in that simple fact.

Romanticism in the Twenty-First Century
Why Romantic Individualism Persists

Why do the romantic gestures discussed in this book keep resurfacing across decades? The attendees at the 1968 Engelbart demo, the hobbyist readers of Ted Nelson in 1970s, the journalists celebrating Steve Jobs as a rebel entrepreneur in the 1980s, the *Wired* readers of the early 1990s, and the open source advocates of the late 1990s all shared an enthusiasm for romantic gestures and interpretations in a way that played a fundamental role in the evolution of computer communication technology and how we use it. What does the history of the internet's development tell us about the persistence and effectivity of romanticism as a social formation?

The Weberian narrative of disenchantment with the modern (with which I began the discussion of romanticism in chapter 2) provides a compelling general sense of the draw of romantic postures and narratives; in Weber's terms, faced with life in the iron cage of modernity, we despair at the lack of enchantment and seek for ways to bring it back. Alan Liu, in *The Laws of Cool*, encapsulates what he calls the "culture of information" with the line, "We work here, but we're cool."[4] This nicely captures that sense of distance people sometimes want to create between themselves—their sense of who they are in the world—and the version of themselves that gets expressed through most forms of work, in which they often feel like a cog in a machine. That distance can serve as the platform from which one can take off in a romantic direction.

But the statement explains the meaning of *cool* in terms of what it is not, by way of exception to what is assumed to be the uncool identity associated with working in a particular place. Similarly, the Weberian approach defines enchantment largely negatively, in terms of what it is against (the lifeworlds of modern instrumental rationality) and explains it in a compensatory manner, not in terms of its positive content. And to the extent that Weber does provide a sense of the positive content of enchantment, he tends to define it as operating nostalgically, as a backwards look towards what it is that we seem to have lost. Yet the technology centered kinds of romanticism discussed here are in the first instance *forward* looking. While there are occasional gestures towards a restoration of certain past conditions like the global village, what internet romanticism has established without a doubt is that romanticism can be constructed within the very technology-centered world that Weber imagined to be enchantment's opposite.[5]

So, how is this pattern best explained? First, it needs be said that in many cases, it was not digital technology itself that transmitted romantic ideas. Romantic tropes, in fact, were largely picked up in *printed* texts. Most of the romantic habits of thought surrounding computers, especially before the mid-1990s, were made available to individuals by way of traditional print (and even *Wired* magazine, at the moment of its greatest impact around 1993–1994, was still largely encountered in print). Just as the sport of Alpine mountain climbing would not be what it is without all the book shelves full of writing about it, the sense that the experience of using computers can be understood as unpredictably thrilling and creative would not exist without all the writing and reading of printed magazine articles, essays, and books about romantic creativity in the field of computing and more generally.

Second, computer countercultural romanticism had a specific history, a cultural context. John Perry Barlow, Ted Nelson, and Stewart Brand had read and created reams of 1960s countercultural literature, as had many of their readers. And, by the late 1970s, computer engineers in general, even those without obvious countercultural proclivities, had encountered enough of these practices that someone inside the Pentagon research establishment like Lynn Conway could describe a giant technological project with gestures towards the informal—"we don't have to form some institute"—and the gestures would be legible.

Third, romanticism was reactive. Romanticism should not be overgeneralized to the spirit of the times in the Hegelian sense, an essence that permeated all aspects of society and culture. It made sense only if it had something to be against, something with which it could be contrasted. In the case of the computer counterculture, it gained traction as a response to other specific modes of thought and their contradictions. When Licklider and Engelbart sought justifications for using computers in an interactive, open-ended way, they were working their way out of a context dominated by instrumental reasoning harnessed to burgeoning corporate and military bureaucratic structures. The dominant logics of systems science and the office automation movement wanted to eliminate the unpredictable, whereas Licklider and Engelbart wanted to explore it. When journalists celebrated Jobs and Wozniak as romantic rebel entrepreneurs, they knowingly did so in the context of a society where entrenched corporate bureaucracies, monotonously uniform national name-brand consumer products, and dry, predictable, profit calculations were the norm; it was precisely the apparent difference from the norm that made the two guys in a garage stories worth telling—to the point where those stories were systematically exaggerated. When *Wired* helped launch both the stock bubble and the internet into everyday life, it arrested our attention precisely because it stood out against a backdrop of routine career paths, political behavior, and capitalist processes. It did not matter much whether or not the Day-Glo graphics and innovative layout of the magazine were readable; it mostly mattered that it stood out from everything else. And when Eric Raymond convinced various corporate chieftains that the best programmers would rather write programs that they were passionate about than programs for which they were well paid, this was made compelling when set against the seemingly intractable Microsoft monopoly and the relatively colorless software it produced.

Computers as Unpredictability Machines

Yet there remains the question of why computers—of all things!—seemed receptive to being articulated with romantic tropes to such effect. Given their initial construction as the embodiment of instrumental reasoning, how could they come to seem to so many a locus of one of instrumental reasoning's primary antagonists?

This is where the compulsive quality of interacting with a computer played a role. Most computer users have had the experience of getting absorbed in web surfing or programming and then finding themselves loosing track of the passage of time, and ending up in a place they had not intended. We have seen how this experience can be assigned any number of different meanings (for example, addiction, exhilaration), and can be (and probably most often is), simply sloughed off as an oddity, as meaningless. Yet, for a significant minority, it does allow for an articulation with two key elements of romanticism: the assertion of unpredictability, and the claim to the distinctness of inner experience.

One of the great strengths of the romantic critique of various rationalisms involves the limits of predictability. The Enlightenment hypothesized that, just as the motions of the planets were discovered to be mathematically predictable by the same rules that made apples fall from trees, so might other aspects of life, such as human behavior. Romanticism's reply is that, at least in the realm of human affairs, this is not so. The romantic thinkers were fascinated by language, art, and history because they are historical and driven by systems internal to themselves, by immanent, necessarily contingent processes that will travel paths that cannot be predicted in advance.

Interactive computers offer the person at the keyboard a world of relatively uncertain outcomes. The results of using a computer are as often as not less predictable than turning on and using, say, a washing machine or a lawn mower. This can be interpreted as a flaw or error (why won't the damn thing do what it should?). Yet there are short steps from noting the unexpected "flaw," to inputting something in response, to a steady, ongoing interaction in which one goes beyond seeking to get the behavior initially intended and goes in unintended directions. Used interactively, computers can become, in a specific way, *un*predictability machines. It is a limited unpredictability, to be sure, more akin to reading a story about a dangerous mountain climbing expedition than to actually being a participant. The safely enclosed experience, the *limited unknowability*, of web surfing or hacking can draw one in and then become articulated with the romantic value of being involved with something beyond the bounds of fully predictable, calculable rationality in which the initial intention is assumed to be fixed. The experience of drifting while interacting with a computer offers an experiential homology to the romantic sense of exploration, an experience of a self-shaping process that unfolds according to its own logic, that cannot be mapped to some external grid.[6]

That homology becomes particularly active socially, however, when it is mapped on to resistance or skepticism towards efforts to predict, rationalize, and control human behavior. As we have seen, many of the first encounters with interactive computers occurred in contexts where bosses or colleagues were proposing to use instrumental logics to manage their way out of human dilemmas: to somehow control the horror of nuclear warfare, to industrialize secretarial work, to calculate one's way out of inner city strife, to win the Vietnam War by computerizing it, to turn schoolchildren into studious and obedient users of electronic encyclopedias, or to resolve the aching political tensions between democracy and the for-profit corporation with tidy committee structures manned by experts. In a life punctuated both by periods of loosing oneself in a machine and regular encounters with misguided instrumental reasoning, the romantic tradition offered a way to justify and celebrate the former while giving voice to one's suspicions about the latter and, most importantly, a way to connect with others with similar views by articulating a shared experience using learned conventions. Finding oneself as a unique expressive individual meant finding others who also liked to think of themselves that way, creating bonds around perceived difference, whether they were dot-commers recommending throwing caution to the wind and taking the plunge into internet start-ups or open source advocates calling for contributors to the Linux kernel. And, as time went on, this alternative discursive universe developed a track record of relative accuracy; already by the early 1980s, Ted Nelson and Stuart Brand had a better track record of predicting the direction of computing than the many managers like Xerox's McColough who imagined computers as tools of prediction and order.

As microcomputers made the experience of interacting with computers widely accessible, more and more people would have the opportunity to map the experience of compulsive interaction onto a romantically inflected interpretation of themselves and the world around them. Someone working with computers, while coming to a dawning realization about the impossibility of one or another example of rationalist overreaching, might then reinterpret the act of computing as something other than instrumental—to see what they were doing as expression, exploration, or art, to see themselves as artist, rebel, or both, and to find communities with similar experiences that would reinforce that interpretation. In a culture which has a nearly two-century habit of celebrating Emerson's dictum that we should understand and trust the self as "that science-baffling star . . . without calculable elements [which is] at once the essence of genius, of virtue, and of life, which we call Spontaneity,"[7] the experience of interacting with a computer, in the context of a reaction against instrumental reasoning turned pathological, could and did work repeatedly as an enabler or reinforcer of romantic constructions of the self.

The Effects of Internet Romanticism

How did this matter? Romanticism did not cause the internet, certainly not by itself. Many very unromantic, traditional human proclivities and economic and political forces have profoundly shaped the innovation and global adoption of internet technology. For example, the routine human desire for and interest in efficient and rich forms of communication accounts for a large part of the long arc of internet development. The capitalist imperative to develop new products—one can attribute it to creative destruction, the tendency towards overcapacity, or the crisis of overproduction, depending on one's theoretical allegiances—certainly has played a key role in continuous technological exploration and development. Direct and indirect state encouragement and cultivation of technological innovation, alongside the state's role as a stabilizer of market relations and regulator of market forces, has also played a crucial role.

Romanticism did help determine when, how, and in what context the internet was adopted and, in turn, the expectations we have for the internet and how it has been integrated into human life. The broad shift from a vision of computers as calculators to computers as communication devices was at the outset driven more by encyclopedic, rationalist visions than romantic ones. Licklider and Engelbart were for the most part rationalists, though their search for ways to find legitimate reasons to develop open-ended interactivity pointed them in new directions. By the 1970s, however, romanticism in its countercultural variant did much to energize, spread, and cement that change. In the 1980s, the romantic respect for informality played a role in enabling the culture of "rough consensus and running code" that shepherded the internet into existence as a widely available, robust network of networks, and in its different, entrepreneurial form it gave legitimacy and energy to the spread and triumph of the "personal" computer that would provide a necessary part of the technological infrastructure for the internet's explosion the following decade. Romanticism was one factor among many, but it seems plausible that romanticism's influence could have provided the edge that allowed the internet to surpass competing technological efforts couched in European corporate liberal efforts, such as Minitel and OSI in the 1980s.

In the 1990s, romanticism in its *Wired* variant was at first more about reception of the internet than about its construction. But, as the 1990s progressed, the *Wired* vision played an essential role in inflating the stock bubble, which in turn caused the huge and rapid influx of capital into internet-related efforts. As foolish as much of the bubble was, it enabled the rapid build-out and adoption of the technology while quickly swamping alternative possible directions for computer communications. From the machines on our desks to telecommunications backbones, massive, society-wide investments in infrastructure, interface technologies, and computer graphics hardware and software were powerfully fueled by stock-bubble energy, driving the internet into our homes and workplaces, providing the internet a material base from which it would launch itself into everyday life across the globe.

But in playing a role in creating the conditions for the internet explosion, romanticism went on to help establish *expectations* of the internet, especially in the United States, that have in turn shaped its further development and integration into life. Our society has been flooded by stories of unpredictable actions by individuals using computers to throw established authorities into disarray: stories of surprising computer-related business start-ups, from Apple and Microsoft around 1980 to Google today; of peculiar digital inventions taking the world by storm; of internet use by political rebels from Jesse Ventura to Howard Dean to Barack Obama; of disruptive events that throw entire industries into disarray, like college students downloading music or uploading videos. We have come to associate the internet with narratives of appealing unpredictability, so much so that we now use those narratives in making choices about regulating and further constructing the internet.[8] We regularly imagine the internet as a space of free exchange beyond regulation, for example, despite all evidence to the contrary (see, for example, China), and, as a consequence, we tolerate quantities of fraud (for example, spam) and pornography that the polity would find unacceptable in broadcasting and telephony. The habit of throwing money at internet-related businesses in rough proportion to their air of rebelliousness persists, even if dampened by memories of the stock collapses and scandals of the early 2000s.[9]

One of the most powerful forces maintaining the internet's open, anarchic character, in sum, is our memory of the all the romantic stories about the internet; those stories taught us to expect the internet to be liberating and unpredictable, and that expectation helps keep it that way.[10] The internet is open, not because of the technology itself or some uniquely democratic potential hidden inside the technology, but because we have narrated it as open and, as a consequence, have embraced and constructed it as open.

The sudden rise of the open source movement works as proof that romantic individualism is not just an epiphenomenon of or apology for the market but in fact has its own relative autonomy. A romantic narrative of organic communities engaging in programming-as-art and individual expression managed to play a key role in shifting the broad discourse surrounding intellectual property in the United States in a way that changed the received wisdoms operating in industry and—gradually—in law. The open source movement did what other critics of neoliberal trends in internet policy making (for example, the Haubens' communitarianism, the civic republicanism of critics like James Boyle, the blistering neo-Marxist critics,[11] or Thomas Hughes's support for the Vannevar Bush school of thought in the form of the military-industrial-university complex) could not: loosen a linchpin of the mainstream dominance of the neoliberal myths about the internet as a triumph of a Lockean marketplace.

In the broad scheme of things, it is too soon to tell whether the internet will turn out to be a truly epochal technology or whether it will simply take its place in the long line of communication technologies that have been elaborating the possibilities of human interconnection across time and space for the last five hundred years. But what will probably stick out when historians look back at the internet's role in the 1990s and 2000s will be the *metaphorical* power of the internet in that decade, the degree to which the perception of the internet as something essentially unpredictable and tied to expressive freedom spilled over into other issues. In the United States at least, one will not be able to understand the larger trajectories of industrial regulation, of antitrust enforcement, of first amendment law, of intellectual property law, of property law itself, without reference to the public's and policy makers' belief that the internet embodied a kind of new energy on the scene that confounded dominant habits of thought. If the internet had been received as simply another embellishment on electronic communication, as something akin to the fax machine or DVD player, events in the worlds of business and government would have played out differently. As often as not, as we have seen, these perceptions were not particularly subtle or even accurate, but they would not have the power they did without the internet as material object functioning as a historically important resource for collectively thinking new thoughts about how to govern society. The connection between romanticism and the internet has not been a relation of blueprint to artifact, but it has been a materially significant one nonetheless.

The Limits of Internet Romantic Individualism

Studying romanticism as a tradition of necessity calls into question romanticism's self image: the claim to be the overthrower of tradition becomes itself a tradition, something learned as much as it is discovered. Law professor Yochai Benkler, discussing the value of internet-based peer-production, once cautioned about what he called "the extravagance of saying this is about freedom."[12] The point is well taken, in large part for empirical reasons; in a country in which large numbers of citizens are faced with the aching lack of nutrition, health care, and other necessities, in which noncitizens and even some citizens are deprived of the most basic protections of the due process of law, in which large concentrations of wealth systematically undermine democratic processes, how important can convenient computer communication really be, especially in a world that already has ubiquitous telephones and photocopiers, cheap video cameras, fax machines, and countless other ways to communicate? It is one thing to argue, say, that requiring telecommunication companies to allow access to competitors' hardware would be reasonable, or efficient, or perhaps even helpful to an open democratic society. It is another to demand, as some have, freedom for the iPhone or to claim that internet regulation boils down to a battle between the forces of freedom and the forces of constraint.[13]

But part of what makes romanticism compelling is precisely its extravagance. When Ted Nelson boldly proclaimed in 1974 that "the purpose of computers is human freedom,"[14] what made the claim compelling then was not that he offered a specific argument about the nature of freedom or democracy or that he described any direct experience with public computing; those were still decades away. Rather, it was precisely the slapdash extremity of the claim, the way it threw all caution to the wind while completely inverting the received wisdom, that caught one's attention. For the small number of individuals with experience with computers at the time, one's susceptibility to the argument came less from logic than from experience—from a pleasure in working with computers, perhaps mixed with some doubts about the boss's misguided managerial fantasies about computers as command-and-control devices. Nelson's rhetoric gave that mix of pleasure and doubt a framework in which one could see one's identity as someone who works with computers in a new light. The humor in Nelson's writing worked to soften its grandiosity, to personalize it—without introducing the complexities of reasoned argument and empirical proof.

The clear appeal and power of such moments of deliberate extravagance is undeniable; in many contexts, simply dismissing it as foolish often will only write one out of participation in the debate. And romantic narratives can sometimes work as a corrective to varieties of bureaucratic grandiosity. It was not the Pentagon that invented the internet, and the internet did not emerge automatically as the result of a few bits of legislation or NSF programs. But the stock-bubble-launching story that Netscape's web browser was revolutionary and that young turk programmer Mark Andreessen was its genius creator worked to obscure important aspects of the situation, such as the merely incremental contribution of Netscape's browser or the nonprofit origins of so much of the internet platform that was in place at the time. Similarly, the idea that open source software is the creation of passionate artist-programmer communities freed from the chains of corporate constraint obscures the fact that this kind of programming activity still needs institutional support—academia, for example, or corporate consortia—not to mention social infrastructures; Linus Torvalds, who programmed the first drafts of the Linux kernel while an undergraduate in Finland, has said he would not have taken the time to get Linux going had he not had the reassurance of a Scandinavian social welfare state providing basic needs while he worked for no income.[15]

So, as a framework for social change, romanticism is, by itself, unsustainable. If romanticism's great strength is its critique of rationalist fantasies of predictability, its blind spots concern social relations and historical context. With its focus on heroic narratives, romanticism obscures the broad social relations that make those "heroic" acts possible. Feminist technology writer Paulina Borsook, in criticizing the enthusiasms that were driving the stock bubble, has described what she called the diaper fallacy:

> making babies, or thinking about making babies . . . is fun. Considering the reality of how many times you will really have to change their diapers (or buy them or wash them or dispose of them or manufacture them or pay for those diapers), is not. . . . It's much more fun to think Grand Abstract Thoughts about (Divine?) Providence providing Prosperity—than to be bothered to think about who wipes the noses and picks up the garbage and absorbs the collateral costs and damage for the outfit.[16]

Borsook's notion of the diaper fallacy works on two levels. On the one hand, diaper changing reminds us of all the unglamorous work needed to sustain human life that is both uncelebrated and that goes largely un- or ill-rewarded, while managers, investors, and other mostly male leaders take all the credit. On the other, it points to a theoretical problem of political economy: the problem of unwaged, reproductive labor that has been so usefully discussed in the wake of the feminist critique of Marx's (and others') labor theory of value. As feminist economists have pointed out for some time now, in a certain sense, capitalism has always been afloat on a large body of unwaged labor; for men to leave the home and go to the factory, whether as managers or laborers, someone has to raise the kids. And this is not a mere mechanical task. Necessary work typically done within the family by women, from diaper changing to homemaking to creating and reproducing a culture and institutions—that is, designing social relations—that nurture the next generation, has for the most part happened outside systems of market exchange.[17]

With this in mind, the fact that crucial work might be done for purposes other than profits, and with attention to structures that facilitate cooperation, is hardly a revelation. Free and open source software production is only surprising in that it emerged in the male-dominated world of engineering against the backdrop of a legal and economic regime that tried to harness a radical individualist free market system of ideas to the horse of high tech. Both the celebrants of the productivity of voluntary peer-to-peer labor[18] and their neo-Marxist critics[19] are perhaps a little too astonished that threads of unwaged work done out of passion or commitment are woven into the fabric of contemporary global capitalism. It is not that software engineers discovered a radically new way to make things with the internet; it is that programmers stumbled upon something that is routine to much of the world but ignored by the dominant ideology, and by wrapping that discovery in romantic clothing and technological triumph—in Byronic mirror shades, so to speak—the programmers were able to penetrate the worlds of law and politics in a way that others had not.

But that success comes with blind spots. Borsook was writing specifically about the mid-1990s moment that fused romantic with radical marketplace principles, but the metaphor is generalizable: romanticism personalizes and in a particular way. On the one hand, it draws attention to colorful individuals while obscuring the role of institutions and broad policy making. The individuals at the center of moments of innovation who for reasons of style or power stand out from those around them—Licklider, Engelbart, Nelson, Jobs, Andreessen, Barlow—get the spotlight, while the important work of the many less colorful individuals who were also involved, like Van Dam, tend to become invisible.

On the other hand, romanticism's focus on spontaneous creation-from-nowhere, its presumption that creativity is its own explanation, and its celebration of the overthrow of tradition, obscures history and social context, such as the social institutions that provided necessary sustenance for developments, like school and health care systems, government funded research, and all the many infrastructural systems that nurture scientific research and technological innovation. It is about the social context that allows people to devote energies to experimenting with technologies, not just about the genius of a young Steve Wozniak or Linus Torvalds. In the end, it is about the tendency towards monopoly in capitalism, not the occasionally arrogant behavior of Microsoft executives. It is about building a society in which exploitive behavior is not rewarded in general, not about relying on heroic entrepreneurs like the founders of Google, who, with great dramatic flare, said that one of their governing principles is "don't be evil." Celebrating the unpredictable and the novel is one thing, but doing so at the expense of an awareness of all the diaper changing that keeps us all alive is another.

Capitalism, Culture, Selves

Yet for all its limits, romanticism keeps resurfacing as an organized force, sometimes with impact. What does this suggest about the nature of capitalist societies in general?

As of this writing, it can be difficult to recall what it was like in the mid-1990s, when, for periods of time, one would be treated as pathetically out of touch if one wondered aloud if the AOL/Time-Warner merger was absurd or expressed doubt that the internet had profoundly changed the rules of economics. These bits of dominant common sense evaporated some time ago, with the roughly simultaneous collapse of the dot com stock bubble and the World Trade Towers in the terrorist attacks of 2001. The ensuing ideological shift allowed these beliefs to seem strange and thus much more amenable to analysis; that historical situation formed the context in which this book was largely conceived and drafted.

Now, another set of changes in dominant modes of thinking is afoot. As this book was in the final stages of completion in the fall of 2008, some of the ideological fabrics that this book has spent much effort accounting for—neoliberal deregulatory and market policies, especially—were torn to shreds. An economic crisis in the U.S. housing mortgage market expanded into the worst global economic collapse since the Great Depression. In response, under first a Republican and then a Democratic presidential administration, the U.S. government has taken a series of actions that would have been unthinkable in the preceding thirty years, such as effectively nationalizing a series of core institutions from banks to major mortgage lenders to insurance and car companies. Almost overnight, Keynesianism rose from the dead and has become, for the moment, the reigning common sense among policy makers inside Washington, DC (and in a surprising number of corporate boardrooms). But as of 2009 it is by no means clear how and by what the old ways of thought will be replaced. We are in the first stages of another shift in the configuration of ideologies, in the relations between ideas and power, but the dust from the collapse of the old is still swirling around us and obscures the outlines of the future. So the decades that are the focus of this book now must be peered at through the refractions of yet another change of dominant weltanschauung.

The events discussed in this book need not be seen as merely of historical interest, however, or merely cautionary tales of past mistakes that we should seek to avoid in the future. Looking back at the dying embers of previous ideological frameworks offers lessons about the relations between culture, politics, and economics that could prove useful no matter what form the future takes.

To start with, the fact that romanticism persisted and at times thrived in association with computers stands as evidence for the claim that culture is not simply a reflection of or separate from economic forces but has its own relatively autonomous role to play in macrostructural events. Culture matters, and not just to itself. Romantic cultural habits were a necessary condition for the largest stock bubble in human history, shaped the design and organization of what is becoming the technological fabric for electronic communication worldwide, and has played a role in the enactment of the legal and felt meanings of freedom in our age.

The stories of internet romanticism, moreover, suggest a particular way in which culture matters; romanticism had its impact by offering forms of selfhood, ways of understanding one's own relation to others, to work, and the world, that can be articulated both with various economic behaviors (for example, programming a user-friendly interface or starting a business) and with macroeconomic philosophies (for example, neoliberal policies or open source policies). By offering a collectively articulate, different sense of oneself while legitimizing doubts about the wisdom of authorities, romantic practices take inchoate lived experiences—for example, the compulsive draw of interactive computing or annoyances with the limits of rationalist plans like office automation—and connects them to a compelling and meaningful system of understanding; pursue one's inner truth, and one might find the satisfactions of self-expression or triumph over the powers that be, over those folks who don't get it. This pattern, we have seen, helped create conditions that shifted the direction of major political discourses and entire economic sectors.[20]

The idea of the market, in a certain sense, then, for all its inadequacies as an empirical description of economic conditions, captures a certain kind of utopian hope—the "Lockean" dream that a crystalline system of property rights will provide us our just rewards and that it's possible to do this in a way that works for all deserving folks in all times in places in a way that transcends history and existing inequalities. While Ayn Rand formally celebrated the abstract, selfish, calculating individual of the utilitarian tradition, it has been argued that she was actually the "last romantic"; the fictional and real characters she celebrated were often artists and architects, expansive and exploratory in culture and ideas, not just in the pursuit of profits.[21] While Rand's easy conflation of expressive romantic individualism with utilitarianism may not matter to experts in neoclassical economic theory, it may help explain why their theories are attractive to the culture at large. The idea of markets is made more attractive when it is harnessed to something other than the profit-maximizing idea of selfhood.

Yet the linkage of markets to romantic selfhood is contingent. It is an articulation, in Stuart Hall's sense, not a logical necessity.[22] We have seen that the articulation of romanticism with capitalism as a whole is flexible; sometimes romantic discourse works in concert with capitalist expansion, as it did in the 1980s, and sometimes romanticism pulls against aspects of capitalism, as the open source movement did in the late 1990s. Framing this process in terms of a binary of resistance versus cooptation grants too much coherence and agency to the abstraction called capitalism; capitalism does not need anything, it is an accumulation of a variety of human actions, not an entity unto itself. It is not outside of human agency. And it overstates the universality of the problem to speak of capitalism's cultural contradictions, as if capitalism simply had two, conflicting needs: the need for us to work hard and save and the need for us to be hedonistic and buy. When Max Weber discerned an elective affinity between colonial Protestant culture and the requirements of emergent eighteenth-century industrial capitalism, he was arguing that it was circumstances peculiar to that time and place that helped get capitalism off and running in North America, not that there was some kind of universal bourgeois subject that capitalism needs in all places at all times. Weber, in fact, left open the question of what other kinds of cultural relations to capitalism developed in later eras. What the story of internet romanticism suggests is that the relationship of romantic constructions to capitalist ones is tangential and historically shifting.

Perhaps the larger point is rather that there is a gap between how many people experience existing systems of production rooted in property relations and what they associate with a meaningful life or with freedom. Market relations as many economists imagine them are not fully livable. Market and property relations provide at best a crude approximation of human desires for things like freedom, justice, and expression; it is only certain circumstances and confluences of events that allow that approximation to make sense. It is not just a morality tale to say that life needs to be meaningful, that it is not enough to be offered merely monetary rewards. Conceived as a whole way of life, as a complete principled system, then, capitalism is unlivable over the long term; something more is needed than the calculated drive for profit maximization, which is why people will seek alternatives or seek to articulate the profit drive with other formations. It seems a safe historical generalization to say that, over time, large numbers of people will articulate and seek out forms of life that offer something more or different, forms that are not always nostalgic or backward-looking, forms that can enthusiastically embrace the latest technologies. The exact modes of that articulation can be hugely consequential.

This does not mean that capitalism is in crisis. It merely means that capitalism and its inner workings are neither inevitable nor perfunctory (in the sense of unthinking or automatic), that capitalism will continue to rely on and be shaped by modes of social organization that are themselves not capitalist, and that how those relations with noncapitalist structures are drawn matter. The quandaries within and struggles over capitalist apparatuses that have accompanied the development of internet technology—struggles over the very ideas of ownership, or law, or the organization of power—tell us that the variety of human desires— desires for wealth, of course, but also desires for respect, justice, expression, community, love, or self-transformation—have at best an awkward and tangential relationship to the forms of selfhood offered by the worlds of money, contracts, property, or the power pyramids of global corporations.

(Re)Discovering the Social

In that awkwardness may lay seeds to social change. We need richer understandings of autonomy in terms of freedom *to*, not just freedom from, working constructions of freedom that understand it as a relationship between people, not simply as a lack of relationship. How to come up with a workable, positive understanding of freedom is of course difficult and uncertain; define positive freedom too abstractly or too strictly, and you define it away into just more constraints or bureaucracy. But the experiences of the development of the internet might give us some clues. For all the radical individualism associated with the internet, it has also occasioned some public rethinking of the nature of democratic social relations.

It should be noted that the argument here is neither that the technology itself is somehow democratic nor that the many moments of horizontal cooperation among programmers and the like amount to some kind of democratic revolution. It is common to speak of all the nascent democratic efforts and hopes around the internet and speak of the democratic potential of the technology. While this way of framing the question usefully allows for human agency—we have a choice, it is implied, whether or not to engage that potential—it still assumes that democracy is somehow *inside* the technology itself. But I have not made that particular dodge. *The Net Effect*, by making the case that cyber rhetoric is not best approached as a set of naïve or pernicious political economic claims but as a rich set of historically embedded cultural discourses with varying political articulations, provides both a more precise picture of what's going on and a little bit better sense of a way forward. My claim is not that the internet is inherently democratic but that, because of a set of historical contingencies, the encounter with the internet has opened up and focused attention on questions of democracy that otherwise rarely enter public discourse.

That encounter has succeeded in enlarging the debate where those of us in critical academia have not. When Lynn Conway cultivated an awareness of how to design methods of design in microchip manufacture involving forms of horizontal collaboration, when the creators of Unix promoted the idea of a programming environment based on shared tools and a consciously cultivated, interacting community, when early internet developers gravitated towards the principle of end-to-end design, when these congealed into the collective habits of "rough consensus and running code" for internet development and governance, there was nothing inherently political about these practices. Bursts of technical innovation had often before been powered by moments of intense cooperation between members of various engineering communities, to various degrees deliberately working around or against established hierarchies. Fred Turner is correct that energetic horizontal collaboration among engineers is neither new nor necessarily the utopian moment that open source romantics sometimes seem to imagine; as Thomas Hughes observed in some detail, the engineers that built the ICBMs in the 1950s also operated in a highly cooperative, informal mode, energized in that case by cold war enthusiasms rather than by any enthusiasm for grassroots democracy.[23] Previous such moments had typically been occasioned by wars (hot or cold) or similar types of crisis. The political valence now associated with what has come to be called network-based peer-to-peer production is contingent; it comes from the historical circumstances in which these particular practices emerged.

Those historical circumstances, then, have been crucial to creating the widely shared feeling that something about the structure of the internet is associated more specifically with the political values of openness and democracy. The temptation to pin those values on the technical structure of the internet is strong. As of this writing, for example, we are faced with a political debate about "net neutrality" in which one side rests its political case heavily on a technological argument about the superior nature of the internet's end-to-end design, where the intermediate pieces of the network are kept as simple and elemental as possible and the control over what goes over the network is left to the devices connected to it.[24] The technical principles upon which the internet was built, we are told, mean that efforts to ration bandwidth—that is, ration flows from within the network according to ability to pay—is, not only antidemocratic, but technically naïve. Democracy, it is thus implied, is technologically superior, and the internet is a superior communication technology because of its democratic characteristics. This is an immensely powerful argument, as it associates democracy with the "hard" values of superior technology instead of mere morality.

Yet, for all its power at this moment, the argument rests on a frail foundation.[25] For the last two decades, the temptation to read political morality tales into the story of the internet's technical success have been almost overwhelming, but they have as often as not dissolved under the weight of experience. The mid-1990s libertarian claims that the internet represented a triumph of the free market clearly ignored all the nonprofit and government-sponsored research that helped create it and the centrality of the often nonproprietary shared standards and protocols that made it work. The oft-repeated slogan "the internet interprets censorship as damage and routes around it"[26] has been disproved by the efforts of, among others, the Chinese government. (In the broadest sense, it is probably the case that a fully internet-connected society might make authoritarianism a little more difficult to maintain, but by no means impossible—something which could also be said about the printing press, the telephone, and the photocopy machine.) The internet is not a technical fix for the principal social and political dilemmas of democracy.

U.S. society is rife with well-organized, respected, and established non- or extracapitalist institutions, from municipal governments to rural electric cooperatives to parental diaper changing. What was historically unique about the work of the network pioneers of the 1970s and 1980s is not that they cooperated or worked outside of proprietary formats but that that particular set of extracapitalist procedures appeared in the heart of an emerging high-technology sector in the absence of a military emergency, associated with various countercultural allures, and at a time when American public discourse was aggressively moving towards a universalizing promarket discourse. The utopian character so many see in the history of internet development is not just about the specifics of internet development; it springs from the fact that the specifics stand in such contrast to the dominant discourses of the times in which they emerged. In contrast with the descendants of Vannevar Bush, the internet pioneers did not proudly spout acronyms or appeal to a sense of national emergency, and at crucial moments they had the wisdom to remain informal ("request for comments," "we don't need to form some institute") and to decline federal funds when it was strategically useful to do so. And in contrast with the "age of the entrepreneur" rhetoric of the mainstream political culture of the 1980s, the internet pioneers knew from experience that exclusively profit-maximizing behavior was not enough to create a thriving network system, that a certain amount of political compromise, a carefully cultivated culture of cooperation, and an enthusiasm for the technology for its own sake were all necessary or at least helpful to getting the internet off the ground. So when the internet took the world by storm in the early 1990s and the likes of Republican leader Newt Gingrich tried to claim the internet as evidence in support of his radical promarket philosophies, activists could look back on the creation of the deep structure of the internet and find compelling object lessons that undercut Gingrich's claims.

The open source movement took the loose outlines of the story of internet development and crystallized them into a dramatic narrative that drew sharp lines between open Gnu/Linux and closed Microsoft. One could make the case that the true success story of open source is the internet itself, not Linux. The internet, however, is more a mix of proprietary and open systems that emerged from a decades-long accumulation of small practical decisions. It involved a lot of diaper changing. In contrast, Linux, coupled to the clever purity of the Gnu General Public License, appeared on the scene at precisely the moment of Microsoft's rise to complete dominance of the operating system market, which was also the historical high-water mark of the drive towards ever more proprietary conditions in the world of intellectual property. If the history of the internet invites a thoughtful, ironic reading of the political economic details, Linux invites a heroic, David versus Goliath narrative.

That narrative, to be sure, has had a powerful effect. We have seen that, while the romantic impulse can blind one to the diaper-changing aspects of life, its iconoclasm can loosen the bonds of existing assumptions and institutional hierarchies and provide the conditions for new forms of social experimentation. The case of the internet has provided compelling models of how to organize collective work. What the pioneers of computer networking rediscovered and institutionalized in the 1970s and 1980s, and what the open source movement seized on and turned into a popular movement in the late 1990s is that horizontal, informal cooperation can be simply effective, and in some circumstances it can be more effective than property relations and market competition. The sometimes nonproprietary, sometimes nonhierarchical ways of the internet are no more and no less anarchist or socialist than city governments or rural electric cooperatives; the simple endurance of city governments and cooperatives—and the internet—belie the claim that for-profit structures are always more efficient, and, by the same token, none of them are inherently utopian or necessarily free from oppression. But the example of the internet's creation has provided a galvanizing set of symbols and social thought-objects that have advanced the discussion of how to do democracy in the broader culture. If one wants to make the case for nonprofit institutional structures, the internet is a valuable rhetorical tool, a starting point for discussion, if not an ending point.

None of which is to say that the specifics of internet development are not still worth learning from. The development of computer networking over the decades has repeatedly provided useful object lessons in how the activity of getting groups to communicate effectively horizontally brings social relations to the front of consciousness; whether the task is something small, like calming the waters on a discussion list, or something large, like carefully steering a fragile technological coalition through the shoals of beltway politics and competing corporate interests, working on networks can encourage an awareness of and concern for social relations of a type that goes beyond crude, top-down command hierarchies. The development of the internet provides distinct examples of the ways in which extracapitalist social relations can emerge neither from business nor from big government but from mixed spaces in between the two. Those relations, furthermore, because they operated at the heart of cutting-edge high technology, need not be seen as somehow compensatory, as reactions to market failures or something that happens at the margins of society; they were at the center of one of the more important technological innovations of the twentieth century. Extracapitalist ways of doing things operate at the center, not just at the margins.

Yet if those lessons are to endure, the discussion of them eventually needs to be taken beyond the romantic individualist narratives that have been so useful in bringing them to broader public attention. Linux is not revolutionary; it is an excellent object lesson that disproves the claim that significant innovation comes *only* from profit incentives. The technology of the internet is not inherently democratic, but interesting and rich experiments in how to do democracy have happened so frequently on the internet that we have come to expect them there and have been building that expectation into its legal regulation and underlying code base to the extent that it is now a tradition. Throughout the history of the internet groups of people have brought various sets of social and political concerns into the discussion of the technology in its formation, a discussion that has since shaped how our society has embraced and continued to develop the internet. Much of our embrace of the internet is driven by longstanding cultural traditions that we have brought to the internet rather than what the internet has brought to us; the internet as a technology is inscribed with tradition at least as much as it marks a break from it. It is not the technology, but how we embraced it, that has made it into the open, loosely democratic institution that it is, into today's favorite model for forward-looking, small *d* democratic practice.

But this is a good thing. It tells us that people—not technology, not big institutions—while going about their lives using and creating technologies, have created some new conditions for democratic hope and experimentation. And, as a result, the internet has occasioned a context in which an ongoing exploration of the meaning of core principles like rights, property, freedom, capitalism, and the social have been made vivid and debated in ways that go well beyond the usual elite modes of discussion. Internet policy making has brought to the fore questions about how we imagine creativity and production in organizing social relations. How does creation happen? What social and legal structures best nurture creativity? Finally, the internet offers important object lessons about the importance and difficulties of imagining social relations and making them available for discussion, lessons that we would do well to learn from.

The internet, in other words, is a socially evocative object.[27] It does not by itself guarantee democracy, but the last several decades of internet evolution offer a set of shared experiences that serve as political object lessons about democracy. Those experiences have played a key role in casting into doubt the certainties of some of the reigning ideas of the last fifty years and widened the range of possibility for democratic debate and action, bringing to the surface political issues that have been dormant since the 1960s or earlier. As a result of this historical experience, the internet's history has become inscribed in its practical character and use. But this efflorescence of openness is not the result of underlying truths about technology (or about progress or humanity) breaking through the crusts of tradition and inequality. It is the result of peculiarities of history and culture, of historical contingencies rather than technological necessities. The larger moral of this story, perhaps, is that democracy is a historical accident worth cultivating.

Notes

INTRODUCTION

1. John Durham Peters, *Speaking into the Air: A History of the Idea of Communication* (University of Chicago Press, 2001), 2.

2. See Mary Douglas, *How Institutions Think* (Syracuse University Press, 1986).

3. See Walter Benjamin, "Theses on the Philosophy of History," in *Illuminations*, ed. Hanna Arendt, trans. Harry Zohn (Harcourt, Brace & World, 1968), 253–64.

4. While this book does consult some primary sources, it is more a work of historical interpretation than of original archival research. The book works at the intersection of science and technology studies, policy studies, and cultural studies. It looks broadly at the social construction of internet technology while also going farther afield than is typical for science and technology studies into broad cultural trends, like romanticism, and it finds the connection between culture and technology construction in certain kinds of law and policy formation.

5. See Thomas Streeter, *Selling the Air: A Critique of the Policy of Commercial Broadcasting in the United States* (University of Chicago Press, 1996). Patrice Flichy's recently translated *The Internet Imaginaire* in a way picks up on the approach of *Selling the Air*. Flichy argues that the various rhetorics surrounding the internet need not to be merely debunked but also understood in terms of their role in the internet's construction, broadly construed. Whereas Flichy's work is quite suggestive, *The Net Effect* develops a much more specific and grounded look at particular points of contact between culture and policy formation and does so using somewhat more focused arguments (for example, about the role of romanticism). See Patrice Flichy, *The Internet Imaginaire*, trans. Liz Carey-Libbrecht (MIT Press, 2007).

6. See Streeter, "'That Deep Romantic Chasm': Libertarianism, Neoliberalism, and the Computer Culture," in *Communication, Citizenship, and Social Policy: Rethinking the Limits of the Welfare State*, ed. Andrew Calabrese and Jean Claude Burgelman (Rowman & Littlefield, 1999), 49–64.

7. This would thus in some sense be a contribution to the type of media history that Lisa Gitelman describes as about "the ways that people experience meaning, how they perceive the world and communicate with each other, and how they distinguish the past and identify culture" (Lisa Gitelman, *Always Already New: Media, History, and the Data of Culture* [MIT Press, 2008], 1).

8. See for example, Ted Friedman, *Electric Dreams* (New York University Press, 2005).

9. Throughout this book, I will occasionally use *Lockean* as a shorthand term to refer to a sense of legitimate property rights derived from an individual's labor as the foundation of a just society, while knowing that this refers to a popular theory of rights that ignores much of what John Locke's actual works were concerned with. For an example of the way his work continues to be used to stand for a particular theory of property rights, see these essays by respected legal authorities: Robert P. Merges, "Locke for the Masses: Property Rights and the Products of Collective Creativity," *SSRN eLibrary*, 5 Jan. 2009, papers.ssrn.com/sol3/papers.cfm?abstract_id=1323408, and William W. Fisher III, "Theories of Intellectual Property," in *New Essays in the Legal and Political Theory of Property*, ed. Stephen R. Munzer (Cambridge University Press, 2001), 168–99.

For the classic critical analysis of Locke's theory of property, see C. B. Macpherson, *The Political Theory of Possessive Individualism Hobbes to Locke* (Oxford University Press, 1965). For a discussion of the limits to this way of reading Locke, see John Dunn, "What Is Living and What Is Dead in the Political Theory of John Locke?" *Interpreting Political Responsibility: Essays 1981–1989* (Princeton University Press, 1990), 9–25.

10. See Erkki Kilpinen, "Memes Versus Signs: On the Use of Meaning Concepts about Nature and Culture," *Semiotica* 2008, 8, no. 171 (2008): 215–37.

11. The starting point for this discussion in the sociology of science is typically Robert K. Merton, "Singletons and Multiples in Scientific Discovery: A Chapter in the Sociology of Science," *Proceedings of the American Philosophical Society* (1961): 470–86.

12. These are iterations of a broader question raised by political scientist Langdon Winner two decades ago in a famous essay titled "Do Artifacts Have Politics?" *The Whale and the Reactor: A Search for Limits in an Age of High Technology* (University of Chicago Press, 1988), 19–39. When he wrote this essay, it was popular in environmental circles to claim that nuclear energy was inherently centralizing, authoritarian even, whereas solar energy would be inherently decentralizing and democratic. Winner concludes that sweeping claims like this are too simple, but, after discussing a variety of case studies in which it could be said that in various ways technologies do seem to have politics, he concludes with what he calls a "both/and position." He wants to look closely at both specific technologies and the social contexts in which they are implemented and, in the process of teasing out the relations between contexts and technologies, to develop a better understanding of what exactly is at stake in the effort to build a more democratic life. Winner's question remains, then, unanswered in full.

13. Among the works in cultural studies of the internet, Fred Turner's *From Counterculture to Cyberculture* is one of the few to take seriously Roy Rosenzwieg's call to explain the seemingly odd confluence in computer culture of 1960s countercultural style and idealism with the cold war military ambitions that drove much of the early research that produced the internet. *The Net Effect* offers further explanation of this confluence, drawing substantially on Turner's work. See Fred Turner, *From Counterculture to Cyberculture: Stewart Brand, the Whole Earth Network, and the Rise of Digital Utopianism* (University of Chicago Press, 2008), and Roy Rosenzweig, "Wizards, Bureaucrats, Warriors, and Hackers: Writing the History of the Internet," *American Historical Review* 103, no. 5 (Dec. 1998): 1530–52.

14. Works that set the standard for the historically oriented study of technologies as socially constructed include Wiebe E. Bijker, *Of Bicycles, Bakelites, and Bulbs: Toward a Theory of Sociotechnical Change* (MIT Press, 1997); *The Social Construction of Technological Systems: New Directions in the Sociology and History of Technology*, ed. Bijker, Thomas P. Hughes, and Trevor Pinch (MIT Press, 1989); and Pinch and Bijker, "The Social Construction of Facts and Artefacts: Or How the Sociology of Science and the Sociology of Technology Might Benefit Each Other," *Social Studies of Science* 14, no. 3 (Aug. 1984): 399–441.

15. For example, the computer industry and the U.S. Department of State squabbled throughout the 1990s over whether or not to restrict the export of microprocessors, which at times resulted in situations where computer game consoles—a machine intended as a toy—were threatened with restrictions because of their potential military applications. See William Glanz, "Clinton Seeks to Ease Computer-Export Rules: GOP Lawmakers Cite Security Concerns," 2 July 1999, *Washington Times*, A1.

16. Donna Haraway, "A Cyborg Manifesto: Science, Technology, and Socialist Feminism in the 1980s," *Simians, Cyborgs, and Women: The Reinvention of Nature* (Routledge, 1990), 164. Richard Ohmann says something similar: "Technology . . . is itself a social process, saturated by the power

relations around it, continually reshaped according to some people's intentions" (Richard Ohmann, "Literacy, Technology, and Monopoly Capital," *College English* 47 [Nov. 1985]: 681), though, I should add that Haraway's full quotation contains a qualification: "Technologies and scientific discourses can be partially understood as formalizations, i.e., as frozen moments of the fluid social interactions constituting them," but that "they should also be viewed as instruments for enforcing meanings."

17. Another way to put this would be to say that the question is, What is society doing to itself when engaging in the set of practices we call the internet?

18. Sherry Turkle, *The Second Self: Computers and the Human Spirit* (Simon and Schuster, 1984).

19. I first came up with this title in "The Net Effect: The Internet and the New White Collar Style," a paper delivered to the Information Technology, and Society Workshop at the School of Social Science of the Institute for Advanced Study, 8–10 June 2001, www.sss.ias.edu/publications/papers/paper14.pdf. It should also be acknowledged that Gil Rodman also independently came up with the title "Net Effect" in Gil B. Rodman, "The Net Effect: The Public's Fear and the Public Sphere," in *Virtual Publics. Policy and Community in an Electronic Age*, ed. Beth E. Kolko (Columbia University Press, 2003), 9–48.

20. Ian Watt, *The Rise of the Novel: Studies in Dafoe, Richardson, and Fielding* (University of California Press, 1957), 60.

21. Christina Dunbar-Hester, "Geeks, Meta-Geeks, and Gender Trouble: Activism, Identity, and Low-Power FM Radio," *Social Studies of Science* 38, 1 Apr. 2008, 206.

22. John Frow, *Time and Commodity Culture* (Oxford University Press, 1997), 187.

23. At any given time and place, it's easier to adopt some forms of personhood than others. The pressures one feels to act as, say, a martyr for Allah, a profit-maximizing businessperson, or a self-sacrificing parent vary according to time, place, and context; they are not determinate in a mechanical way, but are deeply shaped by forces of history and social relations.

24. See E. Gabrielle Coleman and A. Golub, "Hacker Practice: Moral Genres and the Cultural Articulation of Liberalism," *Anthropological Theory* 8, no. 3 (2008): 255.

25. Ellen Ullman, "Come in, CQ: The Body and the Wire," in *Wired Women: Gender and New Realities in Cyberspace*, ed. Lynn Cherny and Elizabeth Reba Weise (Seal Press, 1996), 3–4.

26. See Norris Dickard and Diana Schneider, "The Digital Divide: Where We Are Today: A Status Report on the Digital Divide," *Edutopia: What Works in Public Education*, 1 July 2002, www.edutopia.org/digital-divide-where-we-are-today.

27. See Jane Margolis and Allan Fisher, *Unlocking the Clubhouse: Women in Computing* (MIT Press, 2003). Also see Ellen Spertus, "Why Are There so Few Female Computer Scientists?" 1991, dspace.mit.edu/handle/1721.1/7040; Janet Cottrell, "I'm a Stranger Here Myself: A Consideration of Women in Computing," in *Proceedings of the Twentieth Annual ACM SIGUCCS Conference on User Services* (ACM, 1992), 71–76, portal.acm.org/citation.cfm?doid=143164.143214; and Joel Cooper and Kimberlee D. Weaver, *Gender and Computers* (Lawrence Erlbaum Associates, 2003).

28. I use "we" quite deliberately, not because I think there *is* a larger "we" that was involved in previous decisions, but because a larger "we" *should be* involved in such decisions in the future; I write with the political goal to open such struggles up to the whole of society, to take them beyond the narrow circles to which they have been confined, whether those circles are created out of class privilege or technocratic, utilitarian, or romantic forms of identity.

29. E. P. Thompson, *Making of the English Working Class* (Vintage, 1966), 11.

30. Ibid., 9.

31. See Ullman, *Close to the Machine: Technophilia and Its Discontents* (City Lights Publishers, 2001) and *The Bug* (Anchor, 2004).

1. J. C. R. Licklider, "Computers and Government," in *The Computer Age: A Twenty-Year Review*, ed. Michael L. Dertouzos and Joel Moses (MIT Press, 1979), 126.

2. Quoted in Katie Hafner and Matthew Lyon, *Where Wizards Stay up Late: The Origins of the Internet* (Free Press, 2003), 27.

3. See Turkle, "Video Games and Computer Holding Power," *The Second Self: Computers and the Human Spirit*, 64–92.

4. Hafner and Lyon, *Where Wizards Stay up Late*, 33–34.

5. The actual usefulness of computers interestingly remains a matter of debate among economists, often carried out under the heading of the productivity paradox based on research that finds, in aggregate, the overall contribution of computers to productivity is unclear, and in some cases may be neglible. See, for example, Thomas K. Landauer, *The Trouble with Computers: Usefulness, Usability, and Productivity* (MIT Press, 1996).

6. Carolyn Marvin has documented how, in the late nineteenth century, the introduction of the electric light and the telephone were accompanied by various efforts to explore how these technologies could and should relate to the body and social relations. See Carolyn Marvin, *When Old Technologies Were New: Thinking about Electric Communication in the Late Nineteenth Century* (Oxford University Press, 1990). Similarly, Bryan Pfaffenberger has argued that "to study the processes by which new technologies are made meaningful is to study what is arguably an indispensable component of rapid economic growth and development" (Bryan Pfaffenberger, "The Social Meaning of the Personal Computer: Or, Why the Personal Computer Revolution Was No Revolution," *Anthropological Quarterly* 61 (Jan. 1988): 41.

7. See Rosenzweig, "Wizards, Bureaucrats, Warriors, and Hackers," 1530–52.

8. M. Mitchell Waldrop, *The Dream Machine: J. C. R. Licklider and the Revolution That Made Computing Personal* (Penguin, 2002), 175.

9. Ibid., 176.

10. Paul N. Edwards, *The Closed World: Computers and the Politics of Discourse in Cold War America* (MIT Press, 1997), 266.

11. Ibid., 72.

12. See Merton, "Singletons and Multiples in Scientific Discovery: A Chapter in the Sociology of Science," 470–86.

13. For the many contributors to the development of radio communication, see George Jeffrey Aitken, *Syntony and Spark: Origins of Radio* (John Wiley & Sons, 1976). For a thorough discussion of what the Wright brothers did and did not contribute to the development of the airplane, see Tom D. Crouch, *The Bishop's Boys: A Life of Wilbur and Orville Wright* (W. W. Norton, 1989).

14. For the electric light, for example, in 1878 Joseph Swan received a British patent for an incandescent light bulb based on an evacuated carbon filament lamp almost a year before Edison. See "Sir Joseph Wilson Swan," *Britannica Online Encyclopedia*, www.search.eb.com/eb/article-9070587.

15. See Diana Crane, *Invisible Colleges: Diffusion of Knowledge in Scientific Communities* (University of Chicago Press, 1972); Nicholas C. Mullins, *Theories and Theory Groups in Contemporary American Sociology* (Harper & Row, 1973); and Daryl E. Chubin, *Sociology of Sciences: An Annotated Bibliography on Invisible Colleges* (Garland, 1983).

16. This is sometimes called, following the "strong" program in the sociology of science, the principle of symmetry, where ideas about science or technology are approached at least at the outset neutrally, regardless of whether or not they triumphed in the end. For an example, see Kenneth

Lipartito, "Picturephone and the Information Age: The Social Meaning of Failure," *Technology and Culture* 44 (Jan. 2003): 50–81.

17. See Hafner and Lyon, *Where Wizards Stay up Late*, 10.

18. See ibid., 54–63.

19. See G. Pascal Zachary, *Endless Frontier: Vannevar Bush, Engineer of the American Century* (MIT Press, 1999).

20. United States Office of Scientific Research and Development and Vannevar Bush, *Science, the Endless Frontier* (U.S. Government Printing Office, 1945), chap. 3.2, www.nsf.gov/about/history/vbush1945.htm#ch3.2.

21. There has been an ongoing discussion of the limits of "the linear model"; it assumes a perhaps overly clean linearity from basic research to applied research to development to technological diffusion, which is often said to have originated in Bush's *Science, the Endless Frontier*. See Karl Grandin, Sven Widmalm, and Nina Wormbs, *Science-Industry Nexus: History, Policy, Implications* (Science History Publications, 2004).

22. For an overview, see Ellis W. Hawley, "The Discovery and Study of a 'Corporate Liberalism,'" *Business History Review* 52 (Autumn 1978): 309–20. The theory of corporate liberalism is generally credited to the "revisionist" school of American history associated with William Appleman Williams and his students, like Martin Sklar. See William Appleman Williams, *Contours of American History* (W. W. Norton, 1989), and Martin J. Sklar, *The Corporate Reconstruction of American Capitalism, 1890–1916: The Market, the Law, and Politics* (Cambridge University Press, 1988).

23. See Streeter, "What Is an Advocacy Group, Anyway?" in *Advocacy Groups and the Entertainment Industry*, ed. Michael Suman and Gabriel Rossman (Praeger Publishers, 2000), 77–84.

24. One exception is the academic Left, where those patterns were a principle inspiration for the tradition of critical state theory, which rewrote the standard business history of the twentieth-century United States to emphasize the mutual interdependence of big business and big government. See, for example, *Bringing the State Back In*, Peter B. Evans, Dietrich Rueschemeyer, and Theda Skocpol (Cambridge University Press, 1985).

25. See Hafner and Lyon, *Where Wizards Stay up Late*, 37–38, 89.

26. Se ibid., 86.

27. Thomas P. Hughes, *Rescuing Prometheus* (Pantheon, 1998), 10.

28. Descendants of Gramsci often focus on ways that popular acquiescence to dominant ideas is achieved largely by transmitting or articulating ideas from those with power to those without power. The ideas that become popular "common sense," however, are sometimes only a subset of "ruling" ideas; other modes of thought important to maintaining dominant ways of doing things in various ways are kept out of the limelight instead of made popular—in this case by means of the apparati of expertise.

29. See Hughes, *Rescuing Prometheus*.

30. National Research Council, *Funding a Revolution: Government Support for Computing Research* (National Academies Press, 1999). Online at www.nap.edu/readingroom/books/far/

31. In 1962, for example, Engelbart argued that "the stereotyped image of the computer as only a mathematical instrument is too limiting—essentially, a computer can manipulate any symbol in any describable way. It is not just mathematical or other formal methods that are being considered. Our aim is to give help in manipulating any of the concepts that the individual usefully symbolizes in his work, of which those of mathematical nature comprise but a limited portion in most real-life instances" (Douglas C. Engelbart, "Letter to Vannevar Bush and Program of Human Effectiveness," in *From Memex to Hypertext: Vannevar Bush and the Mind's Machine* [Academic Press Professional, 1991], 239).

32. Andries Van Dam, "Hypertext '87 Keynote Address (Transcript)," *Communications of the ACM*, 1 July 1988, 890.

33. See Jennifer S. Light, *From Warfare to Welfare: Defense Intellectuals and Urban Problems in Cold War America* (Johns Hopkins University Press, 2005).

34. See "The Office of the Future: An In-depth Analysis of How Word Processing Will Reshape the Corporate Office," *Business Week*, 30 June 1975, 56.

35. "Good-bye to Gal Friday? Can You Conceive of a Huge Office with Only a Few Secretaries? IBM Can, and So Can Xerox and a Host of Other Companies," *Forbes*, 15 Dec. 1975, 47.

36. Quoted in "A Market Mostly for the Giants: IBM and Xerox Will Leave Little Room for Other Competitors to Move In," *Business Week*, 30 June 1975, 71.

37. See Kent C. Redmond and Thomas M. Smith, *From Whirlwind to MITRE: The R and D Story of the SAGE Air Defense Computer* (MIT Press, 2000), and Hughes, *Rescuing Prometheus*, 3, esp. chap. 2. Also see Edwards, *The Closed World*, chap. 3. Work on these systems played a role in inspiring 1960s work in interactive computer graphics, particularly Ivan Sutherland's precedent-setting 1963 "Sketchpad." See Ivan Edward Sutherland, "Sketchpad: A Man-Machine Graphical Communication System," *Computer Laboratory Technical Reports: UCAM-CL-TR-574*, 3 Sept. 2003, www.cl.cam.ac.uk/techreports/UCAM-CL-TR-574.html.

38. By the time the first iteration of SAGE was complete in 1963, ICBMs had superseded the airborne bombers that SAGE was designed to detect. See Edwards, *The Closed World*, 37–38.

39. Ibid., 181.

40. See Waldrop, *The Dream Machine*, 178–79.

41. See ibid., 158–83. Rheingold describes the ideas of the piece as "so bold and immense that it would alter not only human history but human evolution, if it proved to be true" (Howard Rheingold, *Tools for Thought: The History and Future of Mind-Expanding Technology* [The MIT Press, 2000], 141).

42. See Edwards, *The Closed World*, 266–67.

43. Licklider et al., "Man-Computer Symbiosis," *IRE Transactions on Human Factors in Electronics* (1960): 4.

44. Ibid., 3.

45. The argument here is not that there was no playfulness elsewhere in the computing world or defense establishment. The makers of nuclear warheads may well have taken pleasure in their work. But that pleasure was not how they justified their work to others. What was unique about Licklider is his construction of a *justification* of something like a playful approach.

46. Licklider, *Libraries of the Future* (MIT Press, 1965).

47. Hafner and Lyon, *Where Wizards Stay up Late*, 36, 34.

48. See Stanley G. Smith and Bruce Arne Sherwood, "Educational Uses of the PLATO Computer System," *Science* 192, n.s., 23 Apr. 1976, 344–52. As one of the reviewers of this book pointed out, laptop touchpads and the iPhone and its imitators have been bringing touch screens to digital communication more generally and may mark the beginning of the decline of the Engelbart-influenced computer interface and a return to Plato-like systems.

49. James M. Nyce and Paul Kahn, *From Memex to Hypertext* (Academic Press, 1991), 237.

50. Waldrop, *The Dream Machine*, 27.

51. Bush, "As We May Think," in *From Memex to Hypertext: Vannevar Bush and the Mind's Machine*, ed. Nyce and Kahn (Academic Press, 1991), 89.

52. Licklider does not cite Bush in "Man-Computer Symbiosis." Waldrop suggests that Licklider had heard of Bush's memex proposal but had not read the article at the time. It may have been Engelbart who revived interest in Vannevar Bush's 1945 "memex" proposal, which is now regularly

treated as part of the canon of modern computing. See Waldrop, *The Dream Machine*, 185. Engelbart reports that he was seized by Bush's dream when he originally read "As We May Think" in 1945 in *The Atlantic Monthly*. It might also be significant, however, that one of the ways Engelbart's discussion of "As We May Think" is known to us is through a grant proposal. Engelbart was asking for money. It could not have hurt his case to associate his then-odd project with Bush's enormous authority and power in the engineering community. So perhaps the notion that Bush's "memex" was a prophetic anticipation of the modern interactive desktop computer may owe as much to Engelbart as to Bush himself. Bardini makes a strong case that Engelbart's indebtedness to Bush is overdrawn and cites other more important sources for Engelbart. See Thierry Bardini, *Bootstrapping: Douglas Engelbart, Coevolution, and the Origins of Personal Computing* (Stanford University Press, 2000). Recent scholarship has revealed the work of Paul Otlet from the 1930s, which in retrospect is more prescient than Bush's memex proposal. See W. Boyd Rayward, "Visions of Xanadu: Paul Otlet (1868–1944) and Hypertext," *Journal of the American Society for Information Science* 45, no. 4 (1994): 235–50, and Alex Wright, "The Web That Time Forgot," *New York Times*, 17 June 2008, www.nytimes.com/2008/06/17/health/17iht-17mund.13760031. html?pagewanted=all.

53. See Bardini, *Bootstrapping*, 33–57. For a discussion of the differences between Whorfian relativist and Chomskyan universalist approaches to language, see George Steiner, "Whorf, Chomsky, and the Student of Literature," *New Literary History* 4 (Autumn 1972): 15–34.

54. Max Horkheimer and Theodor W. Adorno, "The Concept of Enlightenment," *Dialectic of Enlightenment*, trans. John Cumming (Herder and Herder, 1972), 3.

55. See Van Dam, "Hypertext '87 Keynote Address (Transcript)."

56. The Engelbart demo was probably first referred to as the "mother of all demos" in Steven Levy, *Insanely Great: The Life and Times of Macintosh, the Computer That Changed Everything* (Penguin , 2000), 42.

57. See Turner, *From Counterculture to Cyberculture*, 110.

58. See Rheingold, *Tools for Thought*, 188–89.

59. See Clifford Geertz, "Deep Play: Notes on the Balinese Cockfight," *The Interpretation of Cultures* (HarperCollins, 1973), 412–54. Geertz's essay depicted a Balinese cockfight as a type of public gambling ritual involving stakes that exceed any rational utilitarian justification. Engelbart's demo, in its risky, radically new vision of computing, had something of the community intensity that Geertz found in the gambling rituals of cockfights in a village in Bali. Although the experience of computer programming, with its "self-motivating" compulsive absorption, has more than once been compared to the experience of gambling, the analogy to deep play would not be that the demo was like a cockfight but that it embodies a critique of rationalism that is at stake in Geertz's work; by using Bentham's notion of deep play, Geertz observed in an instance of social behavior that what Bentham saw as an exception to the rule could be the rule itself. By showing a case of play with irrationally high stakes to be a deeply meaningful event instead of an odd pathology, Geertz was pointing to the limits of Bentham's—and our—assumptions about human rationality and society.

CHAPTER 2

1. Ralph Waldo Emerson, "Self-Reliance," *Selected Essays, Lectures, and Poems* (Bantam Classics, 1990), 160–61.

2. Stewart Brand, "Spacewar: Fanatic Life and Symbolic Death among the Computer Bums," *Rolling Stone*, 7 Dec. 1972, 50–58.

3. Emerson, "Self-Reliance," 160–61.

4. The terms *romance* and *romanticism* have been applied to the internet and cyberculture by others with notable frequency. Since I used the term in a conference paper in 1997, for example, Wolf used it in the title of his history of *Wired*, and Panzaris's dissertation used it to describe the evolution of computers. Most of these cases use the concept evocatively rather than analytically, however. The exception is Coyne's dense and philosophically oriented *Technoromanticism*. I share Coyne's sense that many critiques of rationalist digital discourses "simply move to a romantic orientation, reworking old ground," as well as his belief that philosophical critiques of the implicit assumptions in digital discourses can help us "understand what is at stake in the enterprise [of building a digital world]." Here I am more concerned, however, with the material effectivity of various digital discourses, their specific functionings in history, than I am with Coyne's approach, which tends to treat them as a set of competing philosophical positions (Richard Coyne, *Technoromanticism* [MIT Press, 2001], 7, 15). See Gary Wolf, *Wired: A Romance* (Random House, 2003), and Georgios Panzaris, "Machines and Romances: The Technical and Narrative Construction of Networked Computing as a General-Purpose Platform, 1960–1995" (PhD diss., Stanford University, 2008).

5. The debates about whether or not certain intellectual figures are properly called romantic— was Goethe classic or romantic? Were Emerson and Thoreau merely interested in British and German romantics, or do they constitute an American version of romanticism?—are based on the assumption that romanticism is best understood in terms of great works and great authors. Friedrich Kittler suggests, not just that the German romanticism should be reperiodized—he reads Goethe's *Faust* as romantic instead of *Klassik*—but that, more broadly, one should understand romanticism, not as a collection of texts or a historical period, but as a way of organizing discourse through structured practices of writing, reading, and relating, that is, as discourse networks. See Friedrich A. Kittler, *Discourse Networks, 1800/1900*, trans. Michael Metteer and Chris Cullens (Stanford University Press, 1992). Max Weber's theory of disenchantment is often said to have been influenced by various romantic strains of thought, especially in terms of the sense of loss associated with rational and intellectualized modernity. See H. C. Greisman, "'Disenchantment of the World': Romanticism, Aesthetics, and Sociological Theory," *British Journal of Sociology* 27 (Dec. 1976): 495–507. Yet Weber's first interest was always relentlessly in the empirical existence of social patterns, not responding to intellectual traditions; what's significant about disenchantment is the broad social trend he was perhaps gropingly trying to describe rather than the fact that in that act of description he sometimes drew on familiar tropes and logics learned from romantic poets and intellectuals.

6. Colin Campbell, *The Romantic Ethic and the Spirit of Modern Consumerism* (Blackwell, 1987). One of the useful aspects of Campbell is that he distinguishes romanticism from the practices of sexual romance and the literature on forbidden love, locating romanticism in the emergence of more generalized, broader definitions of pleasure that may extrapolate from but are not coterminous with the traditions of romantic love.

7. For the concept of culture as a toolkit see Ann Swidler, "Culture in Action: Symbols and Strategies," *American Sociological Review* 51 (Apr. 1986): 273–86.

8. See Turner, *From Counterculture to Cyberculture*.

9. Ibid., 261.

10. See Edwards, *The Closed World*, 3–5.

11. Quoted in "Putting the Office in Place," *Business Week*, 30 June 1975, 56.

12. See Hughes, *Rescuing Prometheus*, 166–88, 189–95.

13. See Light, *From Warfare to Welfare*.

14. See Herbert I. Schiller, *Mass Communications and American Empire* (Beacon Press, 1971).

15. See James W. Carey and John J. Quirk, "The Mythos of the Electronic Revolution," *American Scholar* 39, nos. 1–2 (1970): 219–41, 395–424, and Carey, "A Cultural Approach to Communication," in *Communication as Culture: Essays on Media and Society* (Routledge, 1988), 13–36.

16. See Michael A. Hiltzik, *Dealers of Lightning: Xerox PARC and the Dawn of the Computer Age* (Collins Business, 2000), and Erica Schoenberger, *The Cultural Crisis of the Firm* (Wiley, 1997), 182–202.

17. See Hafner and Lyon, *Where Wizards Stay up Late*, 112–13, 205.

18. See Joseph Weizenbaum, *Computer Power and Human Reason: From Judgement to Calculation* (Freeman, 1976).

19. Ibid., 238–39.

20. For example, "Weizenbaum is perhaps the stereotypical 1960's neo-Luddite" (Richard S. Wallace, "From Eliza to A.L.I.C.E. (A.L.I.C.E. AI Foundation)," www.alicebot.org/articles/wallace/eliza.html).

21. Weizenbaum, *Computer Power and Human Reason*, 2.

22. Ibid., 116.

23. Ibid., 116–17.

24. See Levy, *Hackers: Heroes of the Computer Revolution* (Penguin, 2001), 134.

25. See Edwards, *The Closed World*, 181.

26. See Gregory Bateson, *Steps to an Ecology of Mind: Collected Essays in Anthropology, Psychiatry, Evolution, and Epistemology* (Chandler Press, 1972).

27. Theodore H. Nelson, "Complex Information Processing: A File Structure for the Complex, the Changing and the Indeterminate," in *Proceedings of the 1965 Twentieth National Conference of the ACM* (ACM, 1965), 137, portal.acm.org/citation.cfm?doid=800197.806036

28. Ibid., 136–37.

29. See ibid., 141.

30. Ibid., 136.

31. Ibid.

32. Kathleen Woodward, "Art and Technics: John Cage, Electronics, and World Improvement," *The Myths of Information: Technology and Postindustrial Culture* (Indiana University Press, 1983), 176.

33. See the discussion in Noah Wardrip-Fruin and Nick Montfort, *The New Media Reader* (MIT Press, 2003), 211. Turner discusses the New York art scene and its influence on Stewart Brand; see Turner, *From Counterculture to Cyberculture*, 45–51. Also see Turner, "Romantic Automatism: Art, Technology, and Collaborative Labor in Cold War America," *Journal of Visual Culture* 7, no. 1 (2008): 5–26.

34. See Wardrip-Fruin and Montfort, *The New Media Reader*, 245–46.

35. See Nelson, *Computer Lib/Dream Machines: New Freedoms through Computer Screens—a Minority Report* (Nelson [available from Hugo's Book Service], 1974). *Computer Lib* resists conventional citation. It has two "halves," printed back-to-back in the same volume with each half inverted to the other so that it essentially has two front covers, one titled *Computer Lib* and the flip side titled *Dream Machines: New Freedoms through Computer Screens*. As each half has separate page numbers, citations below will refer to page numbers in *Computer Lib* or *Dream Machines*, as appropriate. The copy used here is described as the First Edition, "NINTH PRINTING, Sept. 1983," and thus, though the copyright listed on both first pages is 1974, a description of events of 1975 like the appearance of the MITS Altair are described, and in Nelson's own biography on page 3 of *Computer Lib* he describes activities through 1977. A heavily revised and retypeset edition was published in 1987 by Tempus Books of Microsoft Press; as some of the more interesting historical material was removed in this revision, citations below are to the first edition unless otherwise noted.

36. Quoted in Wardrip-Fruin and Montfort, *The New Media Reader*, 301.

37. See Nelson, *Computer Lib/Dream Machines* (Microsoft Pr, 1987), 9. I have personally met at least one computer professional who told me the same thing.

38. See Nelson, *Dream Machines*, p. "X" (the second page of a lettered, unnumbered "Special Supplement to the Third Printing, August 1975" that starts *Dream Machines*). The passage predicts that Xerox PARC's innovations—those that would lead to the creation of the Macintosh less than a decade later—will lead Xerox to dominate the computer field. But he concludes with the following prescient parenthetical statement: "The above predictions are based, of course, on the assumption of Xerox management knowing what it's doing. Assumptions of this type in the computer field all too often turn out to be without basis. But we can hope" (Nelson, *Computer Lib*, X).

39. See John J. Anderson, "Dave Tells Ahl—the History of Creative Computing (David Ahl's personal narrative)," *Creative Computing* (Nov. 1984): 74.

40. Wardrip-Fruin and Montfort, *The New Media Reader*, 301.

41. Lest there be any doubt that Nelson was familiar with the *Catalog*, on *Dream Machines*, p. 3, he writes, "of course I'm blatantly imitating, in a way, the wonderful *Whole Earth Catalog* of Stewart Brand." He also cites the *Domebook*, a popular instruction manual on geodesic domes, as inspiration. And on p. 69, *Computer Lib* visually quotes the *Whole Earth Catalog*'s cover (and most famous image) with a full page computer-generated image of earth from space, captioned "The Hole Earth Catalog."

42. Nelson, *Computer Lib*, 3.

43. Presumably, Nelson was referring to the comically inaccurate "A Market Mostly for the Giants: IBM and Xerox Will Leave Little Room for Other Competitors to Move In."

44. Nelson, *Computer Lib*, X.

45. Ibid., 22. Robert Hobbes Zakon's "Hobbes Internet Timeline" attributes the first use of the term *web surfing* to Jean Armour Polly in 1992. See Robert Hobbes Zakon, "Hobbes' Internet Timeline—The Definitive ARPANET and Internet History," www.zakon.org/robert/internet/timeline/.

46. Nelson, *Computer Lib*, 58.

47. Ibid.

48. Nelson, "Keynote Address" presented at the Association of Internet Researchers Fifth Annual Conference, Sussex University, 21 Sept. 2004.

49. See Nelson, *Computer Lib*, 2.

50. See ibid.

51. See ibid., 63.

52. The Amish themselves are not so much opposed to technology in general as they are opposed to forms that create dependency on what they view as the fallen contemporary world. Their break with modern technological forms, in any case, is more radical and differently framed than that of romantics.

53. William Wordsworth, *The Collected Poems of William Wordsworth* (Wordsworth Editions, 1998), 569; Walt Whitman, *The Complete Poems* (Penguin, 2005), 482–83.

54. See Robert Darnton, "Readers Respond to Rousseau: The Fabrication of Romantic Sensitivity," *The Great Cat Massacre and Other Episodes in French Cultural History* (1984): 215–56.

55. Ibid., 228–29.

56. See "HomeLib Issue: BYTE–Issue 13 (September, 1976)," www.devili.iki.fi/library/issue/158.en.html.

57. See Levy, *Hackers*, 154–278. See also Robert X. Cringely, *Accidental Empires: How the Boys of Silicon Valley Make Their Millions, Battle Foreign Competition, and Still Can't Get a Date* (Collins Business, 1996); Paul Sen, "The Triumph of the Nerds: The Rise of Accidental Empires" (PBS, June 1996), www.pbs.org/nerds/; And Martyn Burke, *Pirates of Silicon Valley*, VHS Tape (Turner Home Entertainment, 2000).

58. See Levy, *Hackers*, 220.

59. "IBM Enters 'the Office of the Future,'" *Business Week*, 14 Feb. 1977, 46.

60. With 4KB of RAM, it cost $1298 and $2638 with 48KB of RAM; see the original price list displayed at "Apple II computer," oldcomputers.net/appleii.html.

61. Apple's first disk drive was not available until July 1978, at which point the cost for drives had dropped dramatically.

62. See Victor K. McElheny, "Computer Show: Preview of More Ingenious Models; Computer Show Confirms Trend in Widening Use of Technology," *New York Times*, 16 June 1977, H79.

63. See Richard W. Langer, "Computers Find a Home (Yours)," *New York Times*, 25 Aug. 1977, C48.

64. See Levy, *Hackers*, 265.

65. Ibid., 267.

66. See "TIME Magazine Cover: The Computer, Machine of the Year–Jan. 3, 1983–Person of the Year–Science and Technology–Computers–Machines," www.time.com/time/covers/0,16641,19830103,00.html.

67. See IBM Information Systems Division, Entry Systems Business, "Press Release: Personal Computer Announced by IBM," 12 Aug. 1981), IBM online archives, www-1.ibm.com/ibm/history/documents/pdf/pcpress.pdf.

CHAPTER 3

1. Sen, "The Triumph of the Nerds: The Transcripts, Part III" (PBS, June 1996), www.pbs.org/nerds/part3.html. Sen, "The triumph of the nerds." "Triumph of the Nerds: The Transcripts, Part III," http://www.pbs.org/nerds/part3.html.

2. In an interview Wozniak said, "I get more mention than I deserve. For some reason I get this key position of being one of two people that started the company that started the revolution. Steve and I get a lot of credit, but Mike Markkula was probably more responsible for our early success, and you never hear about him" (Jason Zasky, "Steve Wozniak Interview–Apple Computer–Business–Failure Magazine," *Failure Magazine*, failuremag.com/index.php/feature/article/steve_wozniak_interview/).

3. According to Forbes.com,

Markkula was Chairman of the Board of Apple Computer from January 1977 to May 1983 and from October 1993 to February 1996 and was a director from 1977 to 1997. A founder of Apple, he held a variety of positions there, including President/Chief Executive Officer and Vice President of Marketing. Prior to founding Apple, Mr. Markkula was with Intel Corporation as Marketing Manager, Fairchild Camera and Instrument Corporation as Marketing Manager in the Semiconductor Division, and Hughes Aircraft as a member of the technical staff in the company's research and development laboratory. ("Armas Clifford Markkula Profile–Forbes.com," people.forbes.com/profile/armas-clifford-markkula/28278).

4. Nexis/Lexis full text searches on 15 June 2006 from 1 Jan. 1980–1 Jan. 1990.

5. Quoted in Charles Brown, Jay Hamilton, and James Medoff, *Employers Large and Small* (Harvard University Press, 1990), 1.

6. See "More Young Millionaires, Please: Europe Has Plenty of Entrepreneurs, But Too Few Innovators," *The Economist*, 4 Feb. 1989.

7. See Cindy Skrzycki et al., "Risk Takers," *U.S. News and World Report*, 26 Jan. 1987. Also see "TIME Magazine Cover: Risk Takers–Feb. 15, 1982–Science and Technology–Computers–Apple–Business," www.time.com/time/covers/0,16641,19820215,00.html.

8. See Curtis Hartman, "The Spirit of Independence," *Inc.*, 1 July 1985, 46: "` We have lived through the age of big industry and the age of the giant corporation. But I believe that this is the age of the entrepreneur.'—Ronald Reagan; ` This is America. You can do anything here.'—Ted Turner."

9. Levy, *Hackers*, 258.

10. John Maynard Keynes, *The General Theory Of Employment, Interest,and Money* (Atlantic Publishers & Distributors, 2006), 351.

11. In the early twentieth century, Louis Brandeis, for example, criticized railroads as "unscientifically" run and thus inefficient and underserving of public support (Thomas K. McCraw, *Prophets of Regulation* [Harvard University Press, 1984], 93).

12. For example, David E. Lilienthal, *TVA: Democracy on the March* (Harper & Brothers, 1944).

13. See Eugene Victor Rostow, *Planning for Freedom: The Public Law of American Capitalism* (Yale University Press, 1959).

14. Classics of this school of thought include Richard Posner, *Economic Analysis of Law* (Little Brown, 1974), and R. H. Coase, *The Firm, the Market, and the Law* (University of Chicago Press, 1988).

15. See Daniel Bell, *The Coming of Postindustrial Society* (Basic Books, 1976); Marc Uri Porat, *The Information Economy: Definition and Measurement* (U.S. Department of Commerce, Office of Telecommunications, 1977); and Anthony G. Oettinger, *Elements of Information Resources Policy: Benefits and Burdens in the New World of Library and Other Information Services* (Harvard University, Program on Information Technologies and Public Policy, 1975).

16. See Jean-Francois Lyotard, *The Postmodern Condition: A Report on Knowledge* (University of Minnesota Press, 1984).

17. Bell, *The Coming of Postindustrial Society*, 128.

18. See Frank Webster and Kevin Robbins, *Information Technology: A Luddite Analysis* (Ablex Publishing, 1986), and Nicholas Garnham, "'Information Society' as Theory or Ideology: A Critical Perspective," *Information, Communication and Society* 3, no. 2 (2000): 139.

19. The exceptions are James Beniger, *The Control Revolution: Technological and Economic Origins of the Information Society* (Harvard University Press, 1989), and, late in the game, Manuel Castells, *The Information Age: Economy, Society, and Culture* (Blackwell Pub, 1999).

20. See Zbigniew Brzezinski, "America in the Technetronic Age," *Encounter* 30 (1968): 16-26; James Martin, *Telematic Society: A Challenge for Tomorrow* (Prentice-Hall, 1981); and Stephen H. Lawrence, *Centralization and Decentralization: The Compunications Connection* (Harvard University Center for Information Policy Research, 1983).

21. See Tom Wolfe, "What If He Is Right?" *The Pump House Gang* (Farrar, Straus & Giroux, 1968), 107–33.

22. See Duncan Matthews, *Globalising Intellectual Property Rights* (Routledge, 2002), 15.

23. See Ithiel de Sola Pool, *Technologies of Freedom* (Harvard University Press, 1984).

24. Ibid., 1.

25. See David Kairys, "Freedom of Speech I," in *The Politics of Law: A Progressive Critique*, ed. Kairys (Pantheon, 1982), 237–72.

26. See Brand, *The Media Lab: Inventing the Future at MIT* (Viking Adult, 1987), 214.

27. See Marvin, *When Old Technologies Were New*, 71, and Claude S. Fischer, *America Calling* (University of California Press, 1994), 222–54.

28. See "1982 Brochure for the IBM Personal Computer," digitize.textfiles.com/items/1982-ibm-personal-computer/.

29. Ad produced by Chiatt/Day and directed by Ridley Scott, first broadcast during the 1984 Super Bowl. Malone calls the ad "the biggest splash in the history of advertising" (Michael S. Malone, *Infinite Loop* [Currency/Doubleday, 1999], 274). Also see Sarah R. Stein, "The '1984' Macintosh Ad: Cinematic Icons and Constitutive Rhetoric in the Launch of a New Machine," *Quarterly Journal of Speech* 88, no. 2 (2002): 169–92. Yet most of the attention for the ad came from inside the advertising industry, where it received several awards. The ad, by most accounts, ran nationally only once during the 1984 Super Bowl, and it was not typical even of Apple's advertising, which generally used a much blander, plain folks approach, emphasizing ease of use and functionality. See, for example, the collection of Apple's print ads at *Attached*, www.aresluna.org/attached/computerhistory/ads/international/apple.

30. Edwards discusses *Neuromancer*, *War Games*, and *Terminator* extensively and suggests they had a significant role in the zeitgeist of the 1980s on Edwards, *The Closed World*, 327–45. Also see Friedman, *Electric Dreams*.

31. See Anne Anable, "Computers Made to Feel at Home," *New York Times*, 9 Mar. 1980, C10–11.

32. "A Bright New World of Home Computers; Doing the Homework: Young Experts Tips on Buying; Where to Shop; a Computer Glossary," *New York Times*, 4 June 1981.

33. A Lexis/Nexis search of major U.S. and world publications for the word *spreadsheet* before 1983 turns up 18 hits. The year 1983 turns up 89 hits; 1984, 220 hits; and 1985, 356 hits.

34. See Marbach, "To Each His Own Computer," *Newsweek*, 8 Feb. 1982, 50-56.

35. Ibid.

36. See Mikhail Sergeevich Gorbachev, *Memoirs*, trans. George Peronansky and Tatjana Varsavsky (Doubleday, 1996), 217.

37. See Stuart Hall, "On Postmodernism and Articulation: An Interview with Stuart Hall," ed. Lawrence Grossberg, *Journal of Communication Inquiry* 10, no. 2 (1986): 45.

38. John Sculley, speaking in Sen, "The Triumph of the Nerds: The Transcripts, Part III."

39. Levy, *Hackers*, 7.

40. See Art Kleiner, "A History of *Coevolution Quarterly*," in *News That Stayed News, 1974–1984: Ten Years of Coevolution Quarterly*, ed. Brand and Kleiner (North Point Press, 1986), 371.

41. See ibid., 336–37.

42. See Levy, *Hackers*, 197–98.

CHAPTER 4

1. Steve Crocker, "Interview with Steve Crocker, CEO of Shinkuro, Inc.," *IETF Journal* 2 (Spring 2006), www.isoc.org/tools/blogs/ietfjournal/?p=71#more-71.

2. For example, in the words of Virginia Postrel, editor of *Reason* magazine, "the Net became a model of spontaneous order and decentralized governance—of the way simple, underlying rules can permit enormous creativity and complexity. This dynamic, open-ended vision does not fit easily with the technocratic models that dominate the political world." For libertarians like Postrel, in other words, the supposedly spontaneous order of the internet looked like the market (Declan McCullagh, "The Laugh Is on Gore," *Wired News*, 23 March 1999, www.wired.com/politics/law/news/1999/03/18655/).

3. See Janet Abbate, *Inventing the Internet* (MIT Press, 2000); Paul E. Ceruzzi, *A History of Modern Computing* (MIT Press, 2003); and James Gillies and Robert Cailliau, *How the Web Was Born: The Story of the World Wide Web* (Oxford University Press, USA, 2000).

4. Tarleton Gillespie, "Engineering a Principle: 'End-to-End' in the Design of the Internet," *Social Studies of Science* 36, 1 June 2006, 427–57.

5. Abbate reports that, in the 1970s, contrary to the ARPANET plan, "email quickly became the network's most popular and influential service, surpassing all expectations" (Abbate, *Inventing the Internet*, 107).

6. Licklider and A. Vezza, "Applications of Information Networks," *Proceedings of the IEEE* 66, no. 11 (1978): 1330–46.

7. See Michael Hauben, Ronda Hauben, and Thomas Truscott, *Netizens: On the History and Impact of Usenet and the Internet* (Wiley-IEEE Computer Society Pr, 1997), 254.

8. Lynn Conway, "The MPC Adventures: Experiences with the Generation of VLSI Design and Implementation Methodologies," *Microprocessing and Microprogramming* 10, no. 4 (Nov. 1982): 209–28. Also available online as Conway, "Lynn Conway's VLSI MPC Adventures at Xerox PARC," 19 Jan. 1981, ai.eecs.umich.edu/people/conway/VLSI/MPCAdv/MPCAdv.html.

9. Conway, "Lynn Conway's Career Retrospective," *Lynn Conway Homepage*, ai.eecs.umich.edu/people/conway/RetrospectiveT.html. See especially ai.eecs.umich.edu/people/conway/Retrospective3.html.

10. Dennis Ritchie, "The Evolution of the Unix Time-Sharing System," in *Language Design and Programming Methodology*, 1980, 25–35, dx.doi.org/10.1007/3-540-09745-7_2.

11. Crocker, "Interview with Steve Crocker, CEO of Shinkuro, Inc."

12. M. Doyle, "XEROX, TANG, and DARPA.," *Datamation*, 1 Oct. 1983, 263–64.

13. Conway, "Lynn Conway's Career Retrospective."

14. Doyle, "XEROX, TANG, and DARPA," 263.

15. Alex Roland and Philip Shiman point out that the program failed at its primary mission—creating machine intelligence—but contributed in various ways to some of its original goals and to the general improvement of high performance computing. See Alex Roland and Philip Shiman, *Strategic Computing* (MIT Press, 2002), 325.

16. In 1999, Conway publically revealed on her website that in the 1960s she had undergone sex change surgery, as a consequence was fired by IBM, and was forced to re-create a new life in what she calls "stealth mode" in the 1970s and 1980s, letting her past be known only to close friends and associates. Since making her past public, she has become an advocate and supporter of other transsexual and transgender individuals. The relevance of this to the current discussion is most obviously that the cultural changes of the 1970s inside the invisible college of computer engineers helped make her research possible. Her past was not unknown to those who did a background check so that she could gain security clearance for defense work. That such a person was granted a security clearance to work on cutting-edge military technology in the Pentagon is a sign of cultural changes. But she has also hinted that her experiences as a woman (and transgender person) offered new kinds of insights: "It's hard to put a finger on it, but I was just ever so much more imaginative and creative than before transition—especially noticing that I had vastly improved capabilities at visualizing and mentally simulating complex social interactions" (Conway, "Lynn Conway's Career Retrospective").

17. See Panzaris, "Machines and Romances," 155.

18. Barry Leiner's comments to a Strategic Computing Study Advisory Committee meeting in Arlington, Virginia, on 24 January 1995, are summarized by Roland and Shiman, *Strategic Computing*, 147.

19. See Abbate, *Inventing the Internet*, 184.

20. Abbate writes, "One of the most striking things about the internet in the 1980s was its meteoric growth. In the fall of 1985 about 2000 computers had access to the Internet; by the end of 1987 there were almost 30,000, and by October of 1989 the number had grown to 159,000" (ibid., 186).

21. Ibid., 207.

22. Robert E. Kahn, "Memorial Tribute to Barry Leiner," *D-Lib Magazine* 9, no. 4, www.dlib.org/dlib/april03/04editorial.html.

23. See William J. Broad, "Pentagon Curbing Computer Access: Global Network Split in a Bid to Increase Its Security," *New York Times*, 5 Oct. 1983, A13.

24. Ibid., A13.

25. For the rhetoric of level playing fields, see Streeter, *Selling the Air*, 181–83.

26. For example, see Evelyn Richards, "Semiconductor Industry Wants Nation 'Technology Initiative,'" *Washington Post*, 20 Apr. 1991, B1.

27. See Louise Kehoe, "US Computer Chiefs to Lobby Washington in Battle with Japan," *Financial Times*, 8 June 1989, 6.

28. National Research Council (U.S.). National Research Network Review Committee et al., *Toward a National Research Network* (National Academy Press, 1988). May 23, 1989A, pp. 4-5. Milton Mueller, *Ruling the Root* (MIT Press, 2004), 92.

29. Kahin's biography lists the following as supporters of the IIP: Bellcore, AT&T, IBM, Hughes, Motorola, EDS, Nynex, Digital Equipment, Apple, and Microsoft. It collaborated with a wide range of institutions, including the Global Information Infrastructure Commission, the Coalition for Networked Information, the Freedom Forum, the Annenberg Washington Program, the Library of Congress, the Cross-Industry Working Team, the Computer Systems Policy Project, and the International Telecommunication Union. See Brian Kahin, "Brian Kahin Bio," *Brian Kahin Bio*, www.si.umich.edu/~kahin/bio.html.

30. Office of Science and Technology Policy, *The Federal High Performance Computing Program*, 8 Sept. 1989, pp. 32, 35; quoted in Kahin, "RFC 1192 (rfc1192)—Commercialization of the Internet summary report," www.faqs.org/rfcs/rfc1192.html.

31. High-Performance Computing Act of 1990, 101st Congress, 2d sess., 3 Apr. 1990, S. 1067, title II, sec. 201.

32. Mitch Kapor, "Where Is the Digital Highway Really Heading? The Case for a Jeffersonian Information Policy," *Wired* 1 (July–Aug.1993): 53–59, 94.

33. The earliest reference to "roadkill on the information superhighway" in Lexis/Nexis is David Landis, "Video Dealers Seek a Role in the High-Tech Future," *USA Today*, 12 July 1993, sec. Life, 4D.

34. For example, see William D. Marbach, "The Dazzle of Lasers," *Newsweek*, 3 Jan. 1983, 36: "This year alone, AT&T will install 15,000 miles of glass fibers in commercial systems across the country. Two 'information superhighways' being built of fiber-optic cable will link Boston, New York, Philadelphia and Washington, D.C." The phrase "electronic highway system" dates as far back as 1970; see Ralph Lee Smith, "The Wired Nation," *The Nation*, 18 May 1970, 602.

35. Richard I. Kirkland, Jr., "What the Economy Needs Now," *Fortune*, 16 Dec. 1991, 59.

36. Alan Stewart, "NCF Flexes Its Muscles," *Communications International* (Nov. 1991): 12; quoting Congressman Don Ritter, a member of the U.S. House Science, Space, and Technology Committee.

37. See Joshua Quittner, "Senate OKs $2B for Work on National Computer Net," *New York Newsday*, 12 Sept. 1991, 35.

38. *High-Performance Computing Act of 1990: Report of the Senate Committee on Commerce, Science, and Transportation on S. 1067*, 1990, ERIC, www.eric.ed.gov/ERICWebPortal/contentdelivery/servlet/ERICServlet?accno=ED329226, and statement of Senator Albert Gore, Jr., *Congressional Record* (24 Oct. 1990).

39. See Grandin, Widmalm, and Wormbs, *Science-Industry Nexus.*

40. See en.wikipedia.org/wiki/Vannevar_Bush. Also see Zachary, *Endless Frontier.*

41. Martin Schoffstall, email to com-priv@psi.com, 14 June 1990.

42. Uupsi!njin!cup.portal.com!thinman, email to stev@vax.ftp.com, comint@psi.com, 19 June 1990.

43. The *Summary Report* describes how Stephen Wolff of the NSF outlined the (AUP) that had been governing the NSFNet. He explained: "Under the draft acceptable use policy in effect from 1988 to mid-1990, use of the NSFNET backbone had to support the purpose of ' scientific research and other scholarly activities.' The interim policy promulgated in June 1990 is the same, except that the purpose of the NSFNET is now ' to support research and education in and among academic institutions in the U.S. by access to unique resources and the opportunity for collaborative work.'"

Wolff outlined the distinction between commercialization and privatization of the NSF-NET. The distinction he made is that "commercialization" is "permitting commercial users and providers to access and use Internet facilities and services," while "privatization" is "the elimination of the federal role in providing or subsidizing network services."

44. Allen Leinwand, email to schoff@uu.psi.com (cc: com-priv@psi.com), 15 June 1990.

45. Schoffstall, email to gnu@toad.com (cc: com-priv@psi.com, tcp-ip@nic.ddn.mil), 25 Sept. 1990.

46. "THE TRANSITION: Excerpts from Clinton's Conference on State of the Economy," *New York Times*, 15 Dec 1992, B10; cited in Gordon Cook, "NSFNET ' Privatization' and the Public Interest: Can Misguided Policy Be Corrected?" Jan. 1993, www.cookreport.com/p.index.shtml.

47. See John Schneidawind, "AT&T's Allen Feuds with Gore," *USA Today*, 15 Dec. 1992, 4B.

48. See Kahin, "Interview with Brian Kahin," 24 May 2001.

49. See Richard Wiggins, "Al Gore and the Creation of the Internet," *First Monday* 5 2 Oct. 2000, firstmonday.org/htbin/cgiwrap/bin/ojs/index.php/fm/article/viewArticle/799/708.

50. See Abbate, *Inventing the Internet*, 145.

51. Conway, "The MPC Adventures: Experiences with the Generation of VLSI Design and Implementation Methodologies."

52. The importance of this moment was first suggested to me by a comment made from the audience by Scott Bradner at the conference "Coordination and Administration of the Internet," sponsored by the Kennedy School of Government's Harvard Information Infrastructure Project, Harvard University, 8–10 Sept. 1996.

CHAPTER 5

1. See Dave Farber, "Remarks of John Perry Barlow to the First International Symposium on National Security and National Competitiveness," 21 Feb. 1993, textfiles.tonytee.nl/magazines/SURFPUNK/surf0059.txt.

2. Ibid.

3. See David Toop, "MTV Gets Tangled in the Net," *The Times (London)*, 28 May 1994, 16.

4. See Connie Koenenn, "E-Mail's Mouthpiece: In Just a Year, Wired Magazine Has Become the Guide Down the Information Superhighway," *Los Angeles Times*, 30 Mar. 1994, E1.

5. See Wolf, *Wired–A Romance*, 18–21.

6. See Paul Keegan, "Reality Distortion Field," *Upside.com*, 1 Feb. 1997, www.upside.com/texis/mvm/story.

7. Quoted in ibid.

8. Quoted in ibid.

9. It needs be said because of the seriousness with which investors took statements like that of Netscape founder, Jim Clark: "If the invention at the heart of my first start-up was an internal combustion engine, Mosaic was fire itself" (Jim Clark, *Netscape Time: The Making of the Billion-Dollar Start-Up That Took on Microsoft* [St. Martin's Griffin, 2000], 228). But in fact there were other web browsers before *Mosaic*. In 1992, for example, the X-Windows-based Viola web browser had scrollbars, back and forward buttons in the upper-left corner, a globe icon in the upper right, a URL display, variegated fonts, and of course the ability to move to underlined links through the click of a mouse—just like Mosaic, Netscape, and Internet Explorer. See Ed Krol, *The Whole Internet* (O'Reilly & Associates, 1992), 227–33.

10. Josh Hyatt, "Hyperspace Map: Mosaic Helps Lead Users through Maze of Internet," *Boston Globe*, 29 Mar. 1994, 1.

11. Campbell, *The Romantic Ethic and the Spirit of Modern Consumerism*, 77.

12. Jim Impoco, "Technology Titans Sound off on the Digital Future," *U.S. News and World Report*, 3 May 1993, 62.

13. See Gary Stix, "Domesticating Cyberspace," *Scientific American*, Aug. 1993, 100–110.

14. See ibid., 105.

15. Ibid., 101.

16. Ibid., 110.

17. In the late 1980s, Bradner had been a founder and administrator of NEARNET, a regional TCP/IP network for New England universities, including his employer, Harvard, which had been a participant in the national internet.

18. See Scott Bradner, "Why Now?" *Network World*, 27 Sept. 1993, www.sobco.com/nww/1993/bradner-1993-09-27.html.

19. Ibid.

20. The company was originally called Mosaic Communications but then changed its name to Netscape after complaints from the NCSA about trademark issues. To avoid confusion, the company is referred to here as simply Netscape.

21. In July 1994, at least ten companies had licensed Mosaic from the University of Illinois for commercial development, including the well-connected Spyglass and Spry. See Elizabeth Corcoran, "Mosaic Gives Guided Tour of Internet," *Washington Post*, July 11, 1994, F19.

22. See Clark, *Netscape Time*, 99.

23. See ibid., 100.

24. See ibid., 106.

25. See ibid., 194.

26. Quoted in Keegan, "Reality Distortion Field."

27. See Simson L. Garfinkel, "Is Stallman Stalled? One of the Greatest Programmers Alive Saw a Future Where All Software Was Free. Then Reality Set In.," *Wired* 1 (Apr. 1993): 102.

28. That distinction may belong to Robert Metcalfe, 3Com founder who in June 1994 published a column which began, "Mosaic is doing for the Internet right now what Visicalc, the proverbial killer application, did for the personal computer around 1980" (Bob Metcalfe, "Thanks, NCSA, for Graduating a Few of Your Mosaic Cyberstars," *InfoWorld*, 6 June 1994, 50).

29. See Wolf, "The (Second Phase of the) Revolution Has Begun: Don't Look Now, But Prodigy, AOL, and CompuServe Are All Suddenly Obsolete—and Mosaic Is Well on Its Way to Becoming the World's Standard Interface," *Wired* (Oct. 1994).

30. John Cassidy, *Dot.con: How America Lost Its Mind and Money in the Internet Era* (Harper Perennial, 2003), 96. Also see Cassidy, "The Woman in the Bubble," *The New Yorker*, 26 Apr. 1999, 48.

31. Quoted in Wendy Zellner and Stephanie Anderson Forest, "The Fall of Enron: How Ex-CEO Jeff Skilling's Strategy Grew So Complex That Even His Boss Couldn't Get a Handle on It," *Business Week*, 17 Dec. 2001, 30.

32. See Erik Barnouw, *A Tower of Babel*, vol. 1 of *A History of Broadcasting in the United States* (Oxford University Press, 1966), 232.

33. In 1918, the corporate world imagined radio as a tool exclusively for strategic, point-to-point uses like ship-to-shore and military communication, and it took the radio amateur community (the original hackers of the twentieth century) to discover the pleasures of broadcasting and using radio for entertainment in the 1906–1919 period. When radio became a popular craze in 1920, the corporate world was taken off guard, and it took major strategic reorientations, a new relationship to Madison avenue, and a new regulatory agency and legal constructs to bring things under control again. It was a key moment in the consolidation of Fordism. See Streeter, *Selling the Air*, 84–91.

34. Virginia Postrel, "On the Frontier: From the Wild East of Russian Capitalism to the Evolving Forms of Cyberspace, Esther Dyson Likes the Promise of Unsettled Territory—and the Challenge of Civilizing It," *Reason Magazine*, Oct. 1996, www.reason.com/news/show/30021.html.

35. Christopher Anderson, "The Accidental Superhighway (The Internet Survey)," *The Economist (US)*, 1 July 1995, 4.

36. Ibid., 3.

CHAPTER 6

1. There are many print and electronic versions of Raymond's legendary essay available, which he first presented in May 1997, but the most appropriate would be the version on his own website, complete with a list of changes over the years; see Eric Raymond, "The Cathedral and the Bazaar," *Eric S. Raymond's Home Page*, 21 May 1997, catb.org/~esr/writings/homesteading/cathedral-bazaar/.

2. Nelson, *Computer Lib*, DM 45.

3. Nelson, www.hyperstand.com/Sound/Ted_Report2.html.

4. Nelson, *Literary Machines: The Report on, and of, Project Xanadu Concerning Word Processing, Electronic Publishing, Hypertext, Thinkertoys, Tomorrow's Intellectual Revolution, and Certain Other Topics Including Knowledge, Education and Freedom* (self published, 1983), chap. 2, 38.

5. Ibid., chap. 2, 38.

6. Nelson, Ted_Report2.html.

7. Locke's phrase in the *Second Treatise* was "no one ought to harm another in his life, health, liberty, or possessions" (John Locke, *Second Treatise of Government*, ed. Crawford Brough Macpherson [Hackett Publishing, 1980], 9).

8. Ayn Rand, *Atlas Shrugged* (Dutton Adult, 2005), 106.

9. For an overview of some of the odd twists and turns in early property law in the United States, see Lawrence M. Friedman, *A History of American Law* (Touchstone, 1986), 234–44.

10. See Thomas C. Grey, "The Disintegration of Property," *Property: Nomos XXII* 69 (1980): 69–70.

11. Streeter, *Selling the Air*, 219.

12. See Carol M. Rose, "Crystals and Mud in Property Law," *Stanford Law Review* 40, no. 3 (1987): 577–610.

13. See Fisher, "Stories about Property," *Michigan Law Review* (1996): 1776–98.

14. Jeremy Bentham, *The Works of Jeremy Bentham*, ed. John Bowring (W. Tait, 1843), 501.

15. See Levy, *Hackers*, 229.

16. Fisher divides these into four perspectives that currently dominate theoretical writing about intellectual property: utilitarianism; labor theory; personality theory; and social planning theory. See Fisher, "Theories of Intellectual Property."

17. See Grey, "The Disintegration of Property."

18. See Bernard Edelman, *Ownership of the Image: Elements for a Marxist Theory of Law*, trans. Elizabeth Kingdom (Routledge & Kegan Paul, 1979).

19. Frow, *Time and Commodity Culture*, 187.

20. David Saunders, *Authorship and Copyright* (Taylor & Francis, 1992), 7.

21. See Martha Woodmansee and Peter Jaszi, *The Construction of Authorship* (Duke University Press, 1994).

22. See Michel Foucault, "What Is an Author?" in *Language, Counter-Memory, Practice*, ed. Donald F. Bouchard, trans. Sherry Simon (Cornell University Press, 1980), 113–8.

23. See Jaszi, "Who Cares Who Wrote Shakespeare?" *American University Law Review* 37 (1987): 617, and James D. A. Boyle, "Search for an Author: Shakespeare and the Framers," *American University Law Review* 37 (1987): 625.

24. See Jane M. Gaines, *Contested Culture: The Image, the Voice, and the Law* (The University of North Carolina Press, 1991).

25. For example, see Boyle, "Search for an Author."

26. Rosemary Coombe, *The Cultural Life of Intellectual Properties: Authorship, Appropriation, and the Law* (Duke University Press, 1998).

27. See U.S. Patent and Trademark Office, *Intellectual Property and the National Information Infrastructure: The Report of the Working Group on Intellectual, Property Rights*, Sept. 1995, www.uspto.gov/go/com/doc/ipnii/.

28. John Perry Barlow, "The Economy of Ideas: a Framework for Patents and Copyrights in the Digital Age (Everything you Know about Intellectual Property Is Wrong)," *Wired* 2 (1994): 349.

29. For example, see Streeter, *Selling the Air*, 276.

30. See Lawrence Lessig, "Plastics: Unger and Ackerman on Transformation," *Yale Law Journal* 98 (1988): 1173.

31. A similar intellectual trajectory was adopted by Barack Obama when he became president of the *Harvard Law Review* in 1989, just as Lessig was assuming his first faculty position at the University of Chicago.

32. Lessig, "Social Meaning and Social Norms," *University of Pennsylvania Law Review* 144 (May 1996): 2181.

33. Lessig, "Understanding Changed Readings: Fidelity and Theory," *Stanford Law Review* 47 (1994): 400.

34. See ibid.

35. See Levy, "Lawrence Lessig's Supreme Showdown," *Wired* 10 (Oct. 2002), www.wired.com/wired/archive/10.10/lessig_pr.html.

36. Julian Dibbell, "A Rape in Cyberspace: How an Evil Clown, a Haitian Trickster Spirit, Two Wizards, and a Cast of Dozens Turned a Database into a Society," *The Village Voice* 23, 21 Dec. 1993, 36–42.

37. Levy, "Lawrence Lessig's Supreme Showdown."

38. Ibid.

39. See Christian Zapf and Eben Moglen, "Linguistic Indeterminacy and the Rule of Law: On the Perils of Misunderstanding Wittgenstein," *Georgetown Law Journal* 84 (1995): 485.

40. See David F. Noble, *America by Design* (Oxford University Press, 1979).

41. "Bill Gates once told me that the way to make money in the computer business is by setting de facto standards, by which he meant proprietary standards" (Cringely, "I, Cringely . The Pulpit . Tactics Versus Strategy | PBS," I, Cringely, 2 Sept. 1999, www.pbs.org/cringely/pulpit/1999/pulpit_19990902_000622.html.

42. "The question is not whether Mr. Gates can strain to see even further—the evidence so far suggests not—but whether his skill at making money in the slipstream of other people's technological vision will serve him as well in the next decade as it has for the past two" ("I Have a Dream," *The Economist*, 25 Nov. 1995, 65).

43. Hauben, Hauben, and Truscott, *Netizens*, x.

44. Ibid., 265.

45. See "Linux," *Wikipedia* en.wikipedia.org/wiki/Linux, and Glyn Moody, *Rebel Code: Linux and the Open Source Revolution* (Basic Books, 2002), 190–91. Also see Michael Tiemann, "History of the OSI | Open Source Initiative," www.opensource.org/history.

46. According to Tiemann, "The immediate chain of events that was to lead to the formation of OSI began with the publication of Eric Raymond's paper *The Cathedral and the Bazaar* in 1997" (Tiemann, "History of the OSI | Open Source Initiative").

47. "The ``utility function'' Linux hackers are maximizing is not classically economic but is the intangible of their own ego satisfaction and reputation among other hackers. (One may call their motivation altruistic, but this ignores the fact that altruism is itself a form of ego satisfaction for the altruist). Voluntary cultures that work this way are not actually uncommon; one other in which I have long participated is science fiction fandom, which unlike hackerdom explicitly recognizes `egoboo' (the enhancement of one's reputation among other fans) as the basic drive behind volunteer activity" (Raymond, "The Cathedral and the Bazaar").

48. See William C. Taylor, "Inspired by Work," *Fast Company* 29 (Oct. 1999): 200.

49. Several more of these aphorisms refer to internal states: "*4. If you have the right attitude, interesting problems will find you*," for example, and "*18. To solve an interesting problem, start by finding a problem that is interesting to you.*"

50. The piece does in various ways acknowledge and elaborate the obvious values of cooperation and sharing and thus has to somehow distance itself from the more simplistic forms of romantic individualism. But the idea of creativity is still very much heroic and Promethean. Consider this passage:

> The only way to try for ideas like that is by having lots of ideas—or by having the engineering judgment to take other peoples' good ideas beyond where the originators thought they could go. . . . Andrew Tanenbaum had the original idea to build a simple native Unix for the 386, for use as a teaching tool. Linus Torvalds pushed the Minix concept further than Andrew probably thought it could go -- and it grew into something wonderful. In the same way (though on a smaller scale), I took some ideas by Carl Harris and Harry Hochheiser and pushed them hard. Neither of us was `original' in the romantic way people think is genius. But then, most science and engineering and software development isn't done by original genius, hacker mythology to the contrary. The results were pretty heady stuff all the same—in fact, just the kind of success every hacker lives for! And they meant I would have to set my standards even higher. [Raymond, "The Cathedral and the Bazaar"]

51. See en.wikipedia.org/wiki/Usage_share_of_web_browsers. Also John Borland, "Browser Wars: High Price, Huge Rewards," *ZDNet News & Blogs*, 15 Apr. 2003, news.zdnet.com/2100-3513_22-128738.html.

52. See Tiemann, "History of the OSI/Open Source Initiative."

53. James Gattuso, "Latest C:\spin from the Competitive Enterprise Institute," 8 Dec. 1998, www.politechbot.com/p-00120.html.

54. See Graham Lea, "MS' Ballmer: Linux Is Communism," *The Register*, 31 July 2000, www.theregister.co.uk/2000/07/31/ms_ballmer_linux_is_communism/, and Michael Kanellos, "Gates Taking a Seat in Your Den–CNET News," *CNet News*, 5 Jan. 2005, news.cnet.com/Gates-taking-a-seat-in-your-den/2008-1041_3-5514121.html.

55. Daniel Lyons, "Software: Linux's Hit Men," *Forbes*, 14 Oct. 2003, www.forbes.com/2003/10/14/cz_dl_1014linksys.html.

56. Esther Dyson, "FC: Re: Competitive Enterprise Institute Blasts Open-Source Software," 9 Dec. 1998, www.politechbot.com/p-00128.html.

57. See S. Adler, "The Slashdot Effect: An Analysis of Three Internet Publications," *Linux Gazette* 38 (1999).

58. See Boyle, "Is Subjectivity Possible-The Post-Modern Subject in Legal Theory," *University of Colorado Law Review* 62 (1991): 489.

59. See Boyle, "A Politics of Intellectual Property: Environmentalism for the Net?" *Duke Law Journal* 47, no. 1 (Oct. 1997): 87–116.

60. Lessig, *Code and Other Laws of Cyberspace* (Basic Books, 1999), 7–8.

61. Lessig, "An Information Society: Free or Feudal?" *The Cook Report on the Internet* (Sept. 2003): 102–4.

62. Search conducted online on 10 Mar. 2008.

63. See Milton Mueller, "Info-Communisim? Ownership and Freedom in the Digital Economy," *First Monday* 137 Apr. 2008, firstmonday.org/htbin/cgiwrap/bin/ojs/index.php/fm/article/viewArticle/2058/1956.

64. See "Copyrights: A Radical Rethink," *The Economist*, 23 Jan. 2003, www.economist.com/opinion/displayStory.cfm?story_id=1547223.

65. Nelson, "Welcome to Udanax.com: Enfiladic Hypertext," www.udanax.com/.

66. Kevin Werbach, *Open Spectrum: The New Wireless Paradigm*, working paper, Spectrum Series (New America Foundation, 2002), werbach.com/docs/new_wireless_paradigm.htm.

67. See Micah L. Sifry, "The Rise of Open-Source Politics: Thanks to Web-Savvy Agitators, Insiderism and Elitism Are under Heavy Attack," *The Nation*, 22 Nov. 2004.

68. See "Open Source Journalism," *Wikipedia*, en.wikipedia.org/wiki/Open_source_journalism.

69. See Larry Rohter, "Gilberto Gil Hears the Future, Some Rights Reserved," *The New York Times*, 11 Mar. 2007, www.nytimes.com/2007/03/11/arts/music/11roht.html?_r=1 and "Brazilian Government Invests in Culture of Hip-Hop," *The New York Times*, 14 Mar. 2007, www.nytimes.com/2007/03/14/arts/music/14gil.html.

70. See Karl Polanyi, *The Great Transformation* (Beacon Press, 2001).

71. See Rose, "Property as the Keystone Right," *Notre Dame Law Review* 71 (1995): 329–65.

CONCLUSION

1. Wolf, "The (Second Phase of the) Revolution Has Begun."

2. Among the journalists and pundits who wrote digital technology into romantic narratives, Wolf, along with *Hackers* author Steven Levy, stand apart in their often astute sense of the

ironies and tensions in the lives and events they chronicled. As they spun their romantic stories around the likes of Andreessen and Wozniak, they were able to note many of the tensions surrounding the characters they celebrated, particularly around the principles of for-profit corporate structure and the pleasure in creating things for their own sake. Just as Wolf noted the ambiguity between Andreessen's dragon-slaying stance towards Microsoft and his company's proprietary practices, Levy's *Hackers* at moments caught the tensions between the market's prerogatives and the hacker ethic of open access and free information; it was Levy who called the world's attention to the young Bill Gates's spat with the Homebrew Computer Club over code sharing, and he labeled Richard Stallman "the last of the true hackers" (Levy, *Hackers*, 413), which turned out to be prescient in terms of anticipating the rise of the open source movement nearly a decade before most others did. Wolf and Levy's accounts of the computer culture will still be worth reading long after many of the romantically inflected notions they helped energize have come to seem peculiar or quaint.

3. A necessary component of Obama's successful 2007–2008 campaign, certainly in the primaries, was unprecedented fund raising through small donors coupled to the organization of a collection of highly energized grassroots efforts, both of which not only used the internet but were strategies learned and improved upon from Howard Dean's "internet campaign" for the presidency in 2003. For a nontechnological determinist analysis of this, see *Mousepads, Shoe Leather, and Hope: Lessons from the Howard Dean Campaign for the Future of Internet Politics*, Streeter and Zephyr Teachout (Paradigm Publishers, 2007).

4. Alan Liu, *The Laws of Cool: Knowledge Work and the Culture of Information* (University of Chicago Press, 2004), 76.

5. James Carey once described this as "nostalgia for the future," which is an evocative way of describing the pattern, but it somewhat begs the question of the internal structures of digital romanticism. See Carey, "The History of the Future," in *Communication as Culture*, 152.

6. The argument is not that everyone who has invoked internet romanticism is a computer "addict." The connection between the compulsive quality of computer communication and romanticism should be understood on the level of receptivity. As interactive computers spread through the culture at large, the experience of compulsive interaction became widespread enough that the articulation of those experiences and various kinds of romantic sensibilities became widely accessible.

7. Emerson, "Self-Reliance," 160–61.

8. In U.S. Constitutional law, for example, the Supreme Court has taken it as axiomatic that the internet is uniquely well suited as a medium for wide-open free speech. See John Paul Stevens, "Opinion of the Court," in *Reno v. Aclu*, 521 U.S. 844 (U.S. Supreme Court 1997).

9. See Michael Barbaro and Leslie Walker, "Dot-Coms Get Back in IPO Game," *Washington Post*, 18 Aug. 2004.

10. This idea that the history of the internet—that is, the way we have told stories about how it has appeared—has become embedded in the materiality of the internet itself, bears some resonance with Gitelman's arguments about the inseparability of media history from the media themselves. See Gitelman, *Always Already New*.

11. See Webster and Robbins, *Information Technology*.

12. Yochai Benkler, "Freedom in the Commons—The Emergence of Peer-Production as an Alternative to Markets and Hierarchies, and the Battle over the Institutional Ecosystem in Which They Compete," paper presented at the Symposium on Cybercapitalism, School of Social Science, Institute for Advanced Study, Princeton, N.J., 29–30 Mar. 2001).

13. For example, see "Free the iPhone: Support Wireless Freedom!" http://www.freetheiphone.org/.

14. Nelson, *Computer Lib*, 1.

15. See Terry Gross, "Computer Programmer Linus Torvalds : NPR," interview, Fresh Air from WHYY, 4 June 2001, www.npr.org/templates/story/story.php?storyId=1123917. In this interview, Torvalds said, "For a while I was concerned about money, and I did not like that feeling. . . . I'm convinced that one of the reasons I never had to care about money is that I come from Finland and they have a very strong social network. For example, University was basically free, health care was basically free, so I come from a culture where you kind of don't have to worry about the basics of life. And I think that's one of the reasons why I was able to ignore, psychologically, the commercial aspects of linux. I'd grown up in a culture where commercial aspects maybe aren't as important as they seem to be in the U.S."

16. Paulina Borsook, "The Diaper Fallacy Strikes Again," *Rewired*, 3 Dec 1997, www.paulinaborsook.com/Doco/diaper_fallacy.pdf.

17. For example, see Batya Weinbaum and Amy Bridges, "The Other Side of the Paycheck: Monopoly Capital and the Structure of Consumption," *Monthly Review* 28, no. 3 (1976): 88–103. More recently, see Nancy Folbre, *Who Pays for the Kids?* (Routledge, 1994).

18. See Benkler, *The Wealth of Networks: How Social Production Transforms Markets and Freedom* (Yale University Press, 2007).

19. See Tiziana Terranova, "Free Labor: Producing Culture for the Global Economy," *Social Text* 18, no. 2 (2000): 33–57, and Andrew Ross, *Nice Work If You Can Get It* (New York University Press, 2009).

20. This provides some substance, if you will, to Althusser's claim that ideology is the imaginary relationship of individuals to their real conditions of existence (where *imaginary* is understood to mean mediated by symbols, not false).

21. See Edward Rothstein, "Considering the Last Romantic, Ayn Rand, at 100," *The New York Times*, 2 Feb. 2005, www.nytimes.com/2005/02/02/books/02rand.html. For Rand's romantic tastes in art, see Rand, *The Romantic Manifesto* (Signet, 1971).

22. See Hall, "On Postmodernism and Articulation."

23. See Hughes, *Rescuing Prometheus*, 11, 107.

24. For example, Lessig and McChesney argue, "net neutrality means simply that all like Internet content must be treated alike and move at the same speed over the network. The owners of the Internet's wires cannot discriminate. This is the simple but brilliant 'end-to-end' design of the Internet that has made it such a powerful force for economic and social good: All of the intelligence and control is held by producers and users, not the networks that connect them" (Lessig and Robert W. McChesney, "No Tolls on the Internet," *Washington Post*, 8 June 2006).

25. Panzaris notes what he calls the "rhetorical naturalization" of the internet since the mid-1990s, where the internet is claimed to be the way it is by necessity, and then normative claims are made on that basis. See Panzaris, "Machines and Romances," 1–5.

26. Attributed to John Gilmore in Philip Elmer-Dewitt, David S. Jackson, and Wendy King, "First Nation in Cyberspace," *Time*, 6 Dec. 1993, http://www.time.com/time/magazine/article/0,9171,979768,00.html.

27. This is an echo of Sherry Turkle's phrase, with addition of *socially* to distinguish from her more psychologically oriented sense of "evocative objects." See Turkle, *Evocative Objects: Things We Think With* (MIT Press, 2007).

Index

Feeling: of computer interaction, 7, 17–18, 42, 86, 88, 166, 158; of market, 86–88, 93; romantic, 45, 60, 158. *See also* Experience, felt

Felsenstein, Lee, 64

Feminism, 11, 48, 178

Fischer, Claude S., 201n27

Fisher, William, 142, 189n9, 207n16

Flichy, Patrice, 189n5

Folbre, Nancy, 211n17

Foucault, Michel, 146. *See also* Author-function

Free Software Foundation (FSF), 151, 155

Free speech, 78, 120, 164; in conflict with market, 167; equated with free market, 136–137

Freedom, negative vs. positive, 182–183

Friedman, Ted, 189n8

Friedman, Thomas, 137

Frontier metaphor, 121, 129, 132

Frow, John, 10, 145

Gaines, Jane, 146

Garnham, Nicholas, 200n18

Gates, Bill, 64, 77, 91, 123, 128, 143, 152–153, 161

Geertz, Clifford, 42, 195n59

Gender, 10–13, 48, 178, 202n16

General Magic, 107

General Public License (GPL), 155, 185

Gibson, William, 81, 121, 123

Gilder, George, 98, 132, 136

Gillespie, Tarleton, 202n4

Gillies, James, 94

Gingrich, Newt, 132, 185

Ginsburg, Douglas H., 74

Gitelman, Lisa, 189n7, 210n10

Goethe, 59, 145, 196n5

Google, 175, 179

Gorbachev, Mikhail, 87

Gore, Albert Jr., 106–107, 113, 128; 2000 presidential campaign, 114

Gore, Albert Sr., 108

Gramsci, Antonio, 193n28

Grey, Thomas, 143

Habitus, 68, 123

Hacker ethic, 90, 169

Hackers, 51–52, 89–91

HAL computer, in *2001*, 28, 63, 81, 92

Hall, Stuart, 88, 181

Haraway, Donna, 8, 190n16

Hauben, Michael and Ronda, 94, 117, 154–155, 175

Hawley, Ellis, 193n22

Hayek, Friedrich, 73

Hegel, 143

Herder, J. G., 39

Heritage Foundation, 74

Hiltzik, Michael A. 197n16

History, theory of, 3, 189n4; and contingency, 172, 187; Hegelian, 4, 51; of internet, 20; and media, 189n7; against teleology, 22, 32, 34–35

Homebrew Computer Club, 64, 143

Hoover, Herbert, 24

Horkheimer, Max, 38

Hughes, Thomas, 26–27, 94–95, 175, 183

Hypertext, 6, 21, 54–56, 57, 141; similarity to cross-references, 36

IBM, 28, 30, 48, 57, 122, 157, 159; IBM PC, 68; System 6 computer, 65–66

Ideas, history of, 4–5; from the bottom up, 7

Identity. *See* Selfhood

Individualism, 10, 121; as creative, 45, 51, 145; possessive, 4, 190n9; and poststructuralism, 10; rational, 45; romantic, 9–10, 45–46, 166; utilitarian, 45, 89, 145

Informal language, 53–55, 58, 96–97, 102, 112, 120–121, 171, 185. *See also* Romanticism, and plain language

Information society, theory of, 71, 75–79, 92

Information superhighway, 108, 113, 123–124, 132; earliest uses of term, 203n33, 203n44

Innovation, technological. *See* technological innovation

Intellectual property: as a bundle of rights, 79; changes in, 169; as expression of contradictions of property in general, 139; and information society, 77; and Ted Nelson, 139–140; theories of, 189n9, 207n16. *See also* Property rights

Interactivity, computers, 31; desire for, 39, 41, 172. *See also* Feeling, of computer interaction

International Standards Organization (ISO), 116

About the Author

THOMAS STREETER is Professor of Sociology at the University of Vermont. He is the author of *Selling the Air: A Critique of the Policy of Commercial Broadcasting in the United States*.